FUNDAMENTALS OF
PRACTICAL
ENVIRONMENTALISM

FUNDAMENTALS OF
PRACTICAL
ENVIRONMENTALISM

MARK B. WELDON

CRC Press
Taylor & Francis Group
Boca Raton London New York

CRC Press is an imprint of the
Taylor & Francis Group, an **informa** business

CRC Press
Taylor & Francis Group
6000 Broken Sound Parkway NW, Suite 300
Boca Raton, FL 33487-2742

First issued in paperback 2017

© 2011 by Taylor & Francis Group, LLC
CRC Press is an imprint of Taylor & Francis Group, an Informa business

No claim to original U.S. Government works

Version Date: 20110504

ISBN-13: 978-1-4398-4928-6 (hbk)
ISBN-13: 978-1-138-11477-7 (pbk)

Visit the Taylor & Francis Web site at
http://www.taylorandfrancis.com

and the CRC Press Web site at
http://www.crcpress.com

To Emma, Martin, Lydia, and Charlie

I hope that we will leave our planet in better condition than we found it.

Contents

List of Figures

List of Tables

Preface

Environmentalism is a big word. Not only does it have a lot of letters, but it also has a lot of meanings. Save the baby seals, save the rain forests, save the snail darter, save electricity, save gasoline, save water, save the whales, save the world!

Environmentalism is a new word. As a term, it barely existed a few decades ago. But at its root it is a very old idea. It suggests that we are dependent upon our environment for our survival, and we should protect it as a matter of self-interest. The ancients knew this to be intimately and obviously true. Their survival to the next year, month, or even day was directly tied to their environment. They were beholden to the food it produced and were awed by the dangers it held.

But is this true today, obviously true? No, certainly not, at least for a vast number of us who enjoy a modern, wealthy, and comfortable standard of living. The environment has been subdued and environmentalism has only recently resurfaced as something that merits attention. Its reappearance has been meteoric, however. In a half century, it has grown from an old quaint thought to a generally accepted philosophy fully ingrained within many aspects of our modern society. Simply pay attention to the advertising campaigns of major corporations to see how important it has become to be green, or at least to appear green.

The resurgence of environmentalism has been so rapid that it has become rather thoughtless. Anything green is good, and no cost is too great in the name of protecting the earth. Environmentalism now seems on par with motherhood and love of country, and utterly unassailable.

Environmentalism as theory, or even as presumed truth, is not exactly the same, however, as environmental change. Environmental change means changing ourselves. Change rarely comes easy and rarely comes cheap. Environmental change usually involves hard choices. Is the environment more important than your health, or your job? Is it more important than having enough food to put on the table or having a nice home? Is it more important than the comfort and happiness of your children?

This is not to suggest that environmentalism and life's necessities are mutually exclusive. It is to suggest that they are intimately intertwined, wrapped tightly together like life's own DNA blueprint. Environmentalism can't only be a luxury of the wealthy. It must be actionable for most of us in order to be real, to be more than empty words from the environmental pulpit. It must be sustainable and lead to lasting positive change for ourselves and our descendants. It must fit within our daily lives.

It must be practical.

Acknowledgments

I want to thank my wife Kim for her efforts in providing me the time, space, and encouragement to write this book. Her insight and constructive criticism during the final editing process was invaluable in helping me bring this work to completion.

I also appreciate the time I spent working on my doctorate at the University of Iowa. Many of the ideas within this book germinated in my mind during my time there. Many thanks to the faculty in the departments of environmental engineering and geography for their instruction and council.

A special thanks to Mark Benesh, a very talented local artist who collaborated with me on the original drawings throughout this book. I appreciate his ability to translate my thoughts into pleasing figures that convey a better sense of the concepts within these pages. If you travel the highways and byways of eastern Iowa near our town of Mount Vernon, you will encounter several of Mark's public works in the form of mural paintings on barns and buildings. I feel fortunate to have his art in this book.

1 An Introduction to Practical Environmentalism and the Four Pillars

Life is not so simple anymore. More and more we are being asked to think beyond our borders, beyond ourselves, and to examine the impacts of our choices and our actions. The purpose of this book is to examine just how we do that with regard to environmental issues, both local and global. It is not an easy task, and it is one that certainly deserves more thought and attention than merely believing, or discounting, the latest environmental crisis brought to our attention by whichever media outlet has our ear.

We live in a very unique time. We have begun to care about the global environment. The state and condition of our own backyards has always been important, but now we are concerned about sea levels in the South Pacific, ice thickness in the Arctic, and the amount of jungle in Brazil. We are concerned about these places primarily because we have been trained to be concerned. These three particular places are foreign to most of us and have no direct impact upon our lives. Yet we are aware of them, primarily because of an environmental movement that is only a few decades old but has gained enough momentum to be fully ingrained in our mainstream media, politics, schools, and scientific research.

Perhaps many people throughout history could make the same claim of the uniqueness of their time. Those who lived during the time of Christ, or Mohammed, or Gandhi could certainly make a strong case for the uniqueness of their time. Think about living through the French Revolution, the United States Civil War, or World War II. Can you imagine suffering through the Black Plague in medieval Europe, or the Great Depression of the twentieth century, or witnessing the dawn of the nuclear age erupt over Hiroshima? Clearly, there have been unique periods of significant transformation in human history. There were times of tragedy as well as inspiration, times of suffering as well as perseverance. Our nations and our cultures have been reformed and remade in relatively short periods of time. These transitions have been very profound and have greatly influenced how we live and what we believe. They transformed our ancestors, and through them they have become part of us. Even lacking the direct experience of these events, we remember these times of crisis.

The particular uniqueness of our times that I refer to is certainly more modest. It comes from the realization that the earth is now linked and networked together more tightly than ever before. This is evident in global politics, economics, financial

1

markets, athletics, communications, environmental issues, and so forth. The peoples of the earth have begun the evolution from nationalism to globalism. Like it or not, with voices raised in praise or protest, this transformation has begun with the patience and inevitableness of a spring thaw.

We have established the United Nations, modern humanity's attempt to peacefully create a world government. This body serves as a gathering place for world leaders to debate, discuss, and sometimes even reach consensus on issues that reach across the continents. We have allowed, even encouraged, our national economic systems to become entangled and intertwined with one another. Large multinational corporations operate all over the world. Countries export the products they make at home to distant nations and import many other products from countries far from their borders. In the United States, you can find autos made by companies headquartered in Germany, Great Britain, Italy, Japan, Korea, and Sweden. International companies all around the world collect their profits in dollars, euros, pesos, yen, and many other local currencies. These national currencies are traded and converted via computerized monetary networks quicker than most of us can imagine. Stocks, bonds, currencies, and commodities are traded every moment of every business day from offices and markets all around the world.

Our cultures have begun to merge and overlap as well. Star athletes, actors, and musicians are known throughout the globe. Worldwide sport championships such as the Olympic Games bring many nations together to compete for the right to brag that they are the best in the world. In the United States, where our national championships in baseball, basketball, football, and hockey command the greatest following, there is a growing interest in broader contests, specifically the World Cup of soccer, or football as it is better known throughout the world. Also, whereas our sports teams are generally associated with major cities, there are teams from our Canadian neighbors to the north that are part of these professional sport leagues.

Television programs, movies, and music can flow seamlessly across borders. We in the United States may not realize that our favorite television program might be a remake based on an earlier show produced in the United Kingdom. A musician's accent disappears as she sings her new hit song in concert in Paris, Madrid, or Los Angeles. Actors can adopt the language and mannerisms of their character, and a tough New York movie thug might be played by an Irishman from Dublin. Electronic devices of all sorts now connect themselves through satellites and wireless networks to reach the World Wide Web and beyond. Big news is global news, instantly.

Amid this globalization of culture, economics, and politics, a modern environmentalism has been born and has begun to grow. In some instances, it has taken on a planetary focus as well. We now have global environmental issues. Global warming is probably the most well known, but there are other issues such as ozone holes, species extinction, and air pollution that have the ability to engage multiple countries in an effort to improve the environment. Environmentalism has grown bold enough to demand a seat at the table with all the other important issues that we face.

It is certainly not my intention here to lobby for the environment per se or for my particular view of environmental issues. Rather, I will argue that environmental issues can be framed and viewed in a manner beyond what our media sources communicate to us. We should first develop a logical method to use when we consider environmental issues and when we contemplate changing our behavior and lifestyles in an effort to benefit the environment. My intent is to lobby for the system we use to make our individual decisions.

As such, I will not be giving you the answers to the environmental issues that confront us. I will not preach a specific environmental ethic that requires your conversion to my way of thinking. Instead, I will present the practical environmentalism approach contained within these pages as a valuable tool to improve the way we interact with our environment.

Within this approach, there is a critically important point and distinction to be made. Practical environmentalism is extremely focused on individual decisions. This book was written specifically for individuals trying to figure out what they could or should do to benefit the environment. If you really want to save the planet, read on! If you have a high degree of healthy skepticism, I think you could find something worthwhile within these pages, too. The emphasis of the book will be on you, your attitudes, your beliefs, your values, your knowledge, and so forth. Certainly, I will ask you to become better informed and more knowledgeable, but always within the confines of your personal judgments.

Having said all that, I will still address those big thorny environmental issues that seem to overwhelm us and be totally beyond any one individual's capacity to implement any meaningful change. Even with these issues, however, I will defer first to the individual's perspective. We as individuals will naturally form an opinion on the big issues, and we should not be discouraged from analyzing them as they relate to our own particular circumstance. From this point we can then take part in the larger debate if we wish. It is hoped we will always keep in mind that environmental issues and the actions we consider in response to these issues affect individual people. These issues and our responses could impact not only ourselves but also friends, families, neighbors, and possibly strangers half a world away.

The title of this book, *The Fundamentals of Practical Environmentalism*, hopefully also conveys a sense of the importance of the individual as part of the environmental decision-making process. The "fundamentals" are the basics; they are those things we should understand first and remember best, they are meant for all of us. They should inform and empower us to develop a working knowledge of the subject at hand and then enable us to act upon this new knowledge to produce meaningful results.

This book develops a fundamentally sound and practical approach to our modern form of environmentalism. The "four pillars" mentioned in the title of this first chapter describes a four-part metric specifically developed to be helpful and instructive when considering environmental decisions, actions, and responses. The four pillars are themselves fundamental and practical in that they are easy to understand and apply. They represent a simple and straightforward combination of the critical factors to consider when we attempt to address the environmental dilemmas in our lives.

The four pillars are

1. Environmental Degradation
2. Resource Conservation
3. Economic Progress
4. Personal Benefit

Each pillar is described in detail in Chapters 5 to 7. Suffice it to say for now that these four pillars facilitate creating a balanced perspective when considering environmental issues. In later chapters we will develop a framework to analyze each of the four pillars with respect to the environmental issue at hand and then combine them into a meaningful holistic metric that can guide our actions.

The harm we could potentially inflict upon the environment is not the whole story; our vantage point is obviously bigger than that. We are faced with choices every day that have more to do with economics, education, family, or even recreation than with the health of our environment. Yet, sometimes, these other important matters have environmental impacts, and sometimes, environmental matters can impact the many other facets of our lives. We truly are linked to our environment, and it is becoming clearer and clearer just how strong these ties are.

In subsequent chapters, through real-world examples, I will show how these four pillars can help us decide whether some action, decision, or response will result in the perceived environmental benefit. That is the essence of practical environmentalism. We should be reasonably sure that, when we set out to do something in hopes of protecting the environment, we really do get an environmental benefit, and at a cost we can accept. We have many, many choices of environmental actions we could take, and they have a range of outcomes and a range of costs. It is not all black and white, nor good versus evil. Practical environmentalism is a greenish shade of gray.

Using the four pillars you can create an environmental tally or scorecard that provides a basic estimation of environmental benefit. Throughout this book I will refer to this process and scorecard simply as "The Pillars," admittedly a not-so-subtle reference to the strong foundation we need to make sense of the environmental issues that confront us.

Using the pillars can range from a simple qualitative analysis to a more complex and detailed quantitative examination of any environmental issue and potential response you wish to address. If you are very concerned about endangered species, you can apply the pillars to your decision to support programs or organizations that preserve wildlife habitat. If global warming keeps you up at night, you can use the pillars to rationalize your level of support for instituting a tax on carbon or any other potential remedy for this global concern. You can use the pillars for your own individual actions, and you can use them to assess the actions of large organizations. The pillars work well in both situations, primarily because practical environmentalism recognizes that environmental issues always eventually affect individuals. Smaller issues tend to involve fewer people. Bigger issues can involve the whole world, but the world is made up of individuals, and the sum of our individual actions is our impact on our world.

I will argue that the direction given by the pillars is the critical component in making decisions that will provide environmental benefit. In particular—and pardon the obtuse language—the direction of the sum of the relative measures can give you a quick and simple indication if your decision is a good one. In other words, if each of the four pillars indicates a positive result, then you can have a high degree of confidence that your decision, or proposed action, will result in environmental benefit. If, on the other hand, the pillars show mixed results, some positive and some negative, then there is reason to question whether the perceived benefit is attainable.

Positive results suggest that your environmental action is sustainable. It suggests that there is enough reason and rationale behind the action that it makes sense to go forward based on your own self-interest. This is the very desirable win–win scenario. Do something good for the environment and good for yourself. Negative results suggest sacrifice. Your loss is someone else's gain. These environmental actions often require some legal or political authority to force these activities to happen. In later chapters we will examine both sustainable and sacrificial environmental actions. Using several diverse examples, we will learn to evaluate each of the four pillars and then combine their results in relatively simple fashion to guide our environmental decisions and actions.

The four pillars of environmental degradation, resource conservation, economic progress, and personal benefit were chosen specifically to represent the complexity and interconnectedness of our environment. The first two pillars are both directly associated with the natural environment. The second two are directly associated with people, collectively and individually. Inclusion of both environmental and human factors is critically important as it explicitly recognizes that humanity is part of the environment and exists within it. Inclusion of the human factors also recognizes the practical aspects of daily living in our society. The kids need to be fed, the bills paid, and we need to be at work on time.

The four pillars were specifically chosen and designed to provide a balance between our humanity and the natural world around us. I will present them in the order described previously, but this order does not necessarily carry with it any preference in importance. Environmental degradation is listed first merely because this is book about environmentalism. Within the four pillars it is no more important than the other three. There is no weighting of the pillars that places one above another; they are all equal.

I consider this equality among the pillars to be a fundamentally critical component of Practical Environmentalism. A bias among the pillars would be representative of our all-too-common contentious debate between the determined environmentalist and the devout capitalist. At the extreme, one values the environment above all else while the other cherishes profit and wealth first and foremost. Practical environmentalism is about balance and sustainable progress. It seeks to avoid narrow-minded single-issue approaches to environmentalism. Overemphasizing one part of the metric makes it far too easy to forget the interconnectedness between our environment and ourselves. Such a bias works against the comprehensive solutions that have the best chance for long-term success. Inequality among the pillars is a fundamental contradiction to the intent of this book.

As we have become more and more aware of our environment's complexity, and how interconnected environmentalism has become with the other facets of our lives, we should also recognize that the way we communicate with one another about environmental issues has become very emotional. News headlines and the accompanying graphic images and pictures we see on television or the Internet about the latest environmental catastrophe are meant to grab our attention and touch our hearts. They are meant to evoke feelings of sadness, or anger, or fear. We are shown a seabird covered in oil from the latest spill, or a picture of the ravages of a recent hurricane along with a report predicting an increase in storm severity associated with global climate change.

The environment is marketed to us in advertising campaigns and in political and social debates. Large corporations highlight their environmental responsibility on network television, and small nonprofits send out mass mailings trying to raise funds to address the environmental issue of the day. The environment is the headline, and it is shown to us in a tidy package. The "XYZ" Corporation cares about the environment and therefore we can trust them and their products. The "ABC" environmental advocacy group warns us of the environmental dangers and therefore deserves our contribution to the cause. The appeal to emotion is simple, direct, and often effective. It does not presume that we care much for the details or that we have the time or energy to care enough to become informed.

Practical environmentalism is meant to address environmental issues in a very different manner. Instead of tightly bundling all the details and factors of an environmental issue together and then hiding them behind an emotional façade, the four pillars are meant to deconstruct the issue far enough so that we can see it more clearly. Without some structure to our thinking, it can be very confusing and very difficult to balance the emotion of the issue with the practicality of the response. The four pillars provide us this structure. Faithfully followed, they require us to break down the issue and our considered response into the four main categories of environmental degradation, resource conservation, economic progress, and personal benefit, and then consider each of these areas separately. This approach encourages us to examine each pillar in sufficient detail so that we can clearly understand the impact of our proposed action in each of these critical areas. We then combine the four pillars and reconstruct our proposed response to the broad environmental issue and judge whether our response is appropriate, reasonable, and ultimately beneficial.

Practical environmentalism is not about removing emotion from our environmental decision-making process entirely, but rather it is about including emotion within our thinking in an appropriate manner that does not exclude other motivating factors. Logic should be part of the process, too. So should a sense of fairness. We should allow ourselves to make a judgment as to the efficacy and impacts of our proposed responses. Our immediate gut-level reaction to the environmental warnings and appeals for change are perhaps not the best source for making choices with respect to our environment. As you will see in Chapter 7, the fourth pillar of personal benefit can be largely driven by our emotions if we wish. Our emotions will also impact the way we analyze and judge each of the other three pillars.

Deconstructing environmental issues via the pillars allows us to become more confident so that we understand the issue well enough to make a decent decision. Notice that I did not say the "right decision" because practical environmentalism is not about telling you what is right and what is wrong. The four pillars are meant to help individuals effect change within their own lives that is environmentally beneficial. They are about helping you approach a problem or make a decision that benefits both you and the environment.

Defining environmentalism as being practical is very important to me in writing this book. Beyond having a nice ring to it and hopefully sparking your interest in what it really means, it is important for a couple of reasons. The first is that this book is intended to be useful to real people like you and me in the large and small decisions of our daily lives. This means that we should be able to put these ideas into practice and receive some benefit from our efforts. Enough people taking small yet measurable actions in a positive direction can yield substantial improvements. The second reason is that the term "practical" has the same root as the word "practice," which has several common understandings within the English language that are beneficial to the objectives of this book. Practice means consistent repetitive preparation, which is fundamental to excelling at anything. It is also used in the phrase *the practice of medicine* or *the practice of law*. In these fields it is well understood that the practitioner is dealing with a complex interactive system that cannot be controlled by one individual. It speaks to the humility of our actions and the desire to do no harm as exemplified by the physician's oath.

> I will use those dietary regimens which will benefit my patients according to my greatest ability and judgment, and I will do no harm or injustice to them.

Partial quote translated into English. (National Institutes of Health 2010)

This emphasis on fundamentals, practicality, and the use of the pillars is meant to address a number of issues and factors that tend to cloud our judgment and confuse our ability to take action in a positive way. These issues are often deeply rooted in our environmental decision making. They often exist beneath the surface and hide away from our attempts at enlightened thinking. These issues can be manipulated by those who understand their existence and who are willing to advance their own agendas with little regard for informed collaborative decisions. A partial list of these confounding factors is

- Competing objectives
- Uncertainty
- Measures of success
- The fallacy of prediction
- Assumption of future states
- The problem with percentages
- A history of paranoia
- A crisis mentality

We will describe these factors in more detail in Chapter 4 and discuss how they can impact our environmental decision making. Each represents common pitfalls in our search for environmental answers. These pitfalls are rooted in our human nature and our general preference to accept what we are told instead of thoughtfully questioning the issue in front of us. There is little likelihood of eliminating these factors as some are deeply embedded within us and within our societies. Some are present in our history and some in the way we educate our children.

These factors are not necessarily bad or evil, but rather they are difficult in the sense that they often make it harder to make decent environmental choices. These confounding factors make it too easy to jump to a conclusion. Practical environmentalism through the four pillars is designed to minimize the effects of these confounding factors and lessen our chances to make poor decisions.

The remainder of this book is dedicated to improving the way we make environmental decisions and increasing our ability to take individual action to produce meaningful results. This book is divided into 14 chapters. The first four chapters examine our current state of environmentalism from historical and ethical perspectives. Chapters 5 through 7 highlight the four pillars themselves and describe why they were chosen to be part of the practical environmentalism metric. Chapters 8 through 12 show how the pillars work together to provide practical and meaningful analyses of environmental issues. Many examples are provided and discussed, ranging from small local opportunities to some of the great and contentious issues of our day. Chapter 13 discusses other environmental analysis concepts and tools that have been proposed to give us a better picture of the state of the environment and how we impact it. The four pillars are compared to these other tools to illustrate the relative strengths and weaknesses of each. In conclusion, Chapter 14 offers some final thoughts about the benefits of practical environmentalism along with a top ten list of individual environmental actions that score well with the pillars.

When you reach the end of this book, I hope you will have gained a new understanding of the relationship between yourself and the environment. I hope that your thinking has been challenged and that environmentalism in general has become more meaningful. I hope that you have found some worth and value in examining the logic we use to determine when and how to act on behalf of the environment. I trust you will have been able to recognize this writer's own personal biases in the examples presented and not necessarily accept them as your own. I believe that each of us has the capacity to examine the state and function of the environment through our own personal lenses and to contribute to its betterment.

Over the past several decades, environmentalism has become more complex, and environmental issues have grown larger with potentially far-reaching impacts. In some ways it seems that our modern environmentalism is outpacing our individual understanding of the basic issues. We need some help in regaining our balance.

Modern environmentalism often exists in crisis mode with many confounding and conflicting factors that make it difficult to formulate truly beneficial responses to the many environmental issues we face. The true art of practical environmentalism is the separation of the few crises with the many trivial and not-so-trivial issues that deserve our reasoned consideration but where we are afforded the luxury of time. These are the many issues of our time. If we can train ourselves to handle these many issues well, perhaps we can properly recognize a real crisis before it is upon us. That is the essence of the four pillars.

REFERENCES

National Institutes of Health. 2002. Greek Medicine: The Hippocratic Oath. http://www.nlm .nih.gov/hmd/greek/greek_oath.html (accessed September 21, 2010).

2 A Short History of Environmentalism in the United States

OK, here is the history lesson. It will be short, occupying only a single chapter in this book, and undoubtedly incomplete. While I think history is a very interesting subject and relevant to the environmental issues of today, I am certainly not a historian. Rather than being a comprehensive historical review, it is included here to help us see how much things have changed over the past few decades. I have deliberately focused on a few historical events that I consider to be instructive in making sense of our modern environmentalism. I hope it serves as a good reference point for where we are and where we are going, or where we should be going. The past is important, for it informs us of our beginnings, which are always linked to our future.

I write from the perspective of a North American and a citizen of the United States. I have traveled some outside the borders of my home country and have thoroughly enjoyed the countries and cultures I had the good fortune to experience. While those travels certainly do not entitle nor empower me to write about the history of environmentalism in these places, I have noticed that we are far more similar than we are different. In general we all work to support ourselves and our families. We appreciate our culture and love our country. We all make choices every day based on personal values and economic constraints. We love our children and fully realize that life goes on longer than we do. We know that our legacy will live on through our descendants, and we imagine what their lives might be like. We hope that they will be comfortable and happy.

While the American experience is what I am most familiar with and most comfortable discussing from a historical perspective, I submit that our experience in the United States is, at a minimum, relevant to the experiences of other areas of the world. The American experience is probably most similar to that in other parts of the world with similar economic conditions, but its relevance does not come necessarily from being "wealthy" or "western" as much as it comes from demonstrating the tension between industrial development and the natural environment. Its relevance also comes from the involvement of government and many private organizations in changing the environmental landscape.

Focusing on the history of environmentalism in America is certainly not an endorsement of America's current environmental ethic as being preferred. Throughout the globe, there is a wide variety of environmental approaches and outlooks. There is much debate and disagreement on the major environmental issues of our time, both within and between the countries of the world. The history of

American environmentalism serves merely as a decent starting point in our trek to understand just how we might be able to make better environmental choices.

The modern environmental movement in the United States really came into being during the 1960s. This is the beginning of the environmental movement as we know it now. Before then, concern for the environment was primarily a concern for perceived shortages of natural resources or a concern for the loss of beautiful areas of wilderness. These concerns were borne out in the conservation and preservation movements of the late 1800s and early 1900s in the United States. The famous figures of this era were people like George Perkins Marsh, Gifford Pinchot, Theodore Roosevelt, John Muir, Aldo Leopold, Henry David Thoreau, and Ansel Adams. Both movements placed a great value on the natural world in stark contrast to the prevailing thoughts and values of the day, and both promoted active involvement in managing natural areas.

Conservationists and preservationists differed in their interpretation of the value of nature. Conservationists assumed a very utilitarian view of nature and appreciated the goods and services nature could provide. Timber from forests, fish from the sea, and rich topsoil for the production of food crops were the beneficial products of the natural world. The conservationist would argue that we must be careful not to over-tax the environment and lessen its capability to produce these goods and services.

Preservationists saw beauty in the natural environment and assumed a very aesthetic view of the world around them. Nature was restorative. It nourished the spirit and satisfied the soul. It somehow allowed humanity to reconnect with what is truly important. At its core, preservationism was spiritual.

To the preservationist, the greatest value of nature was nature itself left in its undisturbed state. Preservationists in these early days of environmentalism did not speak directly to the benefits of biodiversity, habitat protection, or ecosystem function. Their arguments, however, for the preservation of nature were well aligned with these more modern notions. Their pleas were defined more by emotional arguments of the irreplaceable qualities of these wild, beautiful places on our earth.

In America, both conservationists and preservationists focused much of their attention on the western United States. This was very practical and sensible as it was the part of the country that was considered relatively undeveloped at the time. What it lacked in population, form, and structure it certainly made up for in beauty, grandeur, and vast open spaces. The West was still wild and sometimes hostile. Vast formidable mountain ranges and great expansive prairies had prevented easy passage from the east and discouraged all except the hardiest American souls from migrating toward the setting sun. Native American tribes had not always been hospitable to the white-skinned settlers intent on occupying their lands. The weather was harsh and at times very unforgiving, especially with few towns and neighbors nearby to count on for assistance. The western lands were up for grabs for those brave enough and strong enough to make a claim.

In the midst of the nineteenth century, railroads began to pierce through this untamed wilderness, and allowed transit and passage from the civilized east to the up-and-coming cities and towns along the Pacific Coast of America. The railroads represented a drastic improvement in the ease of travel between the two American coasts. More and more people had access to the western lands as it became much easier to travel west and experience firsthand the beauty and the possibilities. It became much easier to think of the West as a new economic frontier and attempt to exploit and harvest its resources. Western and eastern began to meld into one great country in the minds of Americans. This transportation revolution would eventually grant the conservationists and preservationists a stage to play out their ideologies.

Conservationists were interested in protecting western lands from overexploitation. To them the vast forests and grasslands should be managed for sustainable yields of timber and forage. To them, the great rivers could be dammed to provide cheap power and large water reservoirs.

The preservationists, however, were interested in protecting western lands from extractive economic development in any form. Leave the trees standing, let the wild animals graze as they have always done, and let the rivers run free. Let humanity come to these beautiful places and observe and appreciate the natural worth of the wild places.

Out of these protectionist, but somewhat contrary, movements came an interesting combination of governmental actions and authority over the western lands. The US government began to nationalize large tracts of land into what we now know as national parks and national forests. The national parks were formed in the preservationist model while the national forests followed the conservationist ethic. National parks were to protect nature from development, whereas national forests were to manage exploitation of the resources contained within them.

The beginnings of both the US Forest Service and the US National Park Service can be traced back to a series of laws enacted over a few decades in the late 1800s and early 1900s. The Forest Reserve Act of 1891, the Organic Administration Act of 1897, and the Transfer Act of 1905 were all important in defining authority to place forest area under governmental control and to establish the governmental organization to carry out this conservation mission. The Antiquities Act of 1906 and the Organic Act of 1916 performed similar functions for preserving land as national parks.

As is often the case, these governmental legislative activities were a response to cultural and political forces already happening to conserve and preserve critically important public areas. For example, George Perkins Marsh wrote about the importance of forests within the broader ecological and hydrological cycles in 1864 (Marsh 1864), 27 years ahead of the creation of the Forest Reserve Act and the first national forest reserve around Yellowstone in 1891 (Williams 2005).

While Yellowstone is officially credited as the first national park established in 1872, Hot Springs Reservation in Arkansas was formed well before then in 1832 (National Park Service 2005). Even four decades earlier, the National Mall in Washington, DC, was established in 1790 (National Park Service 2005), and all of these predate the formal creation of the National Park Service in 1916 (National Park Service 2006).

The creation of national forests and parks mirrored the debate between conservationists and preservationists. The movements had spurred government involvement and eventually established formal agencies within the US government. It is interesting to note that the US Forest Service became part of the US Department of Agriculture and that the US Park Service became part of the US Department of the Interior, a reasonably accurate reflection of their respective conservation and preservation ideologies.

As mentioned previously, the creation and establishment of national forests and national parks in the United States began almost exclusively in the West. The first 17 national forests were created in the West. President Benjamin Harrison created the Yellowstone Park Timberland Reserve in March 1891, now part of the Shoshone and Bridger–Teton National Forests in Wyoming. President Harrison added 14 more national forest reserves in the states of Alaska, Arizona, California, Colorado, Oregon, and Alaska by the end of his term in the spring of 1893. The next US president, Grover Cleveland, added two more forest reserves in Oregon in 1893 (Williams 2005). Similarly, excluding the National Mall in the Capital, the first 30 national parks were all west of the Mississippi. It was not until 1916 that an eastern national park was formed, the Acadia National Park in Maine (National Park Service 2005).

This appointment of western lands was basically satisfactory to both conservationists and preservationists. Major tracts of land were being protected in many of the western states, some as working forests and some as nature preserves. In just a few decades, more than 40 areas were claimed by the US government for protected status. Both groups could claim their fair share of these accomplishments. Occasionally, though, there were clashes—some ideological, certainly, but some real battles, too—over particular pieces of real estate. Perhaps the most famous and instructive of these clashes between the conservationist and preservationist ideologies was the fight for Hetch Hetchy.

Hetch Hetchy is a relatively small valley within the Yosemite National Park in northern California. It is located approximately 160 miles east of San Francisco. In the early 1900s it was a battleground for the classic conservation–preservation struggle over how best to use nature's resources. Robert W. Righter's book *The Battle over Hetch Hetchy: America's Most Controversial Dam and the Birth of Modern Environmentalism* (Righter 2005), grants us a very thorough view of this struggle that lasted over a decade and involved a cast of characters from across the nation. Most of what is summarized here comes from Righter's work.

The city of San Francisco wanted to dam the Tuolumne River that ran through the Hetch Hetchy Valley and create a reservoir to provide water to the city. This action would flood much of the valley and certainly change the aesthetic landscape, and perhaps disrupt or destroy the natural ecological function of the valley.

Hetch Hetchy was an undeniably beautiful valley. Upstream areas of the Tuolumne River were referred to as "the Grand Canyon of the Tuolumne." Waterfalls sprouted from high granite walls, and ancient meadows carpeted the narrow valley floor. Black Oak trees thrived in the meadows and provided a food staple for the Native Americans that inhabited the valley. It was a unique component of the Yosemite National Park and considered second only to the Yosemite Valley itself in natural wonder. It possessed a beauty that could enthrall the nature lover. It was also beautiful, however, to the engineers from San Francisco. The tall mountains that ringed the valley drew close together where the Tuolumne River exited the valley. At the river outlet, the mountains rose up in a near-perfect V shape with the river at the point. These close and steep sloping natural walls could be joined with concrete and steel to dam and restrain vast amounts of water in the valley behind. While it certainly would not be easy, it was possible. The building of great dams is never easy, and site selection usually hinges on what is the most possible. The engineers could not have

FIGURE 2.1 Artistic rendering of an ideal dam site. (Illustrated by Mark Benesh. With permission.)

asked for a better place to build their dam. Figure 2.1 shows an artist's rendering of this most desirable V-shaped topographical situation of a river running between two mountains.

Hetch Hetchy was a monumental environmental battle in its day. It can be reasonably argued that the skirmishes are not completely over even today as the vanquished refuse to surrender their ideals of the ultimate and best purpose for Hetch Hetchy. It was an epic struggle, lasting some 30 years and ending on the floor of the US Congress in 1913 for final resolution (The Sierra Club 2010).

The battle for Hetch Hetchy began quietly in the early 1870s. John Muir, the famous naturalist and later the first president of the Sierra Club, visited the valley in 1871 and wrote about it in 1873. A few years later the city of San Francisco began to consider the Hetch Hetchy valley as a potential site for a dam and reservoir to supply drinking water to the city. In 1890 the Yosemite National Park was founded, which included the Hetch Hetchy valley. Also in that same year, the mayor of San Francisco, James Phelan, made a formal proposal to dam the Tuolumne River and flood the valley. A preservationist private organization, the Sierra Club, soon rose up to oppose the City's plans. The Sierra Club was founded in 1892 to protect the Sierra Nevada mountain chain, starting with Hetch Hetchy.

A 20-year struggle ensued, pitting the Sierra Club against the city of San Francisco. This struggle outlasted several US presidents and their administrations. It was intensely political. National parks were supposed to be protected and off limits to major economic development. The proposed damming of the Tuolumne would permanently and drastically alter a significant part of the second US national park. An act of Congress would be required, and you do not easily get such an act without a large dose of lobbying and debate.

The political fight between the city and the Sierra Club was not confined to an ideological debate. It was in reality a contentious lobbying effort designed to convince those in authority to side with one of the combatants over the other. It involved US presidents and their secretaries of the interior. It involved prestigious engineers producing lengthy reports on the viability and necessity of creating a reservoir for the people of San Francisco. It involved letter-writing campaigns and accusations of graft and hidden motive. Both sides reached across the country to find allies with voice and reputation to aid their cause.

This fight was heavily influenced by money. San Francisco had plenty of financial resources, and at that time the Sierra Club had few. San Francisco could afford to hire a full-time staff to wage this war. The Sierra Club relied on the passion and compassion of their volunteer members and friends. San Francisco could afford to employ well-known and well-regarded experts in the fields of engineering and hydrology to explain why damming the Hetch Hetchy was necessary. The Sierra Club would counter with experts of their own, but they never quite matched the political impact of the City's campaign. The conservationists eventually won the battle for Hetch Hetchy and were allowed by the US government to construct their grand dam.

Money granted San Francisco the advantage in Washington. It also represented a serious set of blinders to the vision and thinking of city staff and leadership. The City had invested a lot of money in the Hetch Hetchy solution, and this very investment created a momentum both powerful and limiting. It created a single-minded focus within city leadership that consistently, strongly, and ultimately effectively promoted damming Hetch Hetchy as the optimum solution. Yet it also became very difficult for the City to seriously consider alternatives to Hetch Hetchy. In fact there were viable options for developing other water supplies for the city. These alternatives had been identified by several groups, including the US Geological Survey, the Spring Valley Water Company (the private water utility serving the city), and, of course, the Sierra Club. But monetary capital as well as political capital had already been committed to the valley in Yosemite. Indecision, reconsideration, and delay usually do not play well for politicians' images. The populace wanted strong decisive men to lead their city, did they not?

It is also interesting to note that this conservation-versus-preservation struggle was not quite how we might picture it today. It was not about saving nature from humanity as much as it was about the particular way that humanity would undoubtedly encroach upon nature. Both sides favored tourism of the Hetch Hetchy valley and used tourism to bolster their arguments.

Conservationists saw an attractive lake sitting behind a magnificent dam with new roads reaching from the Bay Area east into the wilderness to allow the city dwellers to enjoy the newfound waters. Preservationists envisioned hotels and businesses dotting the valley floor and catering to hikers, campers, and fishermen intent on revitalizing themselves through the experience of nature's wonderful valley. It really was not a struggle about whether they should develop the valley or leave it alone. It was a struggle about just how drastically they would alter the valley. Would they adorn it with buildings and businesses, or would they drown it under tons of water. Which alternative would really be the most beautiful?

This question of beauty may have been one of the key turning points of the political struggle. The preservationists of the day recognized the raw beauty of the valley and were willing and eager to facilitate the journeys of many visitors to experience it. They did not seem to share our more modern notion that development of any kind spoils the pure experience of nature. The conservationists also saw beauty in the valley but one that was a combination of nature's magnificence and humanity's ingenuity. Was a long mountain lake any less beautiful because of the dam that created it? Were the waters somehow less desirable because humans had a hand in their congregation? Were the waterfalls less grand because they fell into a lake rather than a river? The congressmen that were asked to render judgment were presented two options, both vigorously defended as beautiful but with one meeting a great human need for the city of San Francisco.

It is somewhat ironic that perhaps the most significant turning point of the struggle came from nature herself. During the extended debate and political posturing, the city of San Francisco suffered a major earthquake and fire. The 1906 quake caused a three-day conflagration that caused extensive damage. It is estimated that 3,000 people died (Hansen and Condon 1989) and 28,000 buildings were destroyed (US Department of Commerce 1972), mostly by fire. Financial losses were estimated to be 400 million dollars (US Department of Commerce 1972). The city's water system proved inadequate to fight the fire, and its resolve to establish a plentiful, reliable, and economical source of water was stoked even further.

The 1906 natural disaster provided San Francisco the leverage it needed to ultimately prevail in the conquest of the Hetch Hetchy valley. The death and destruction in the city in 1906 were unquestionably horrific, and the potential environmental damage of damming a small river in a small valley probably paled in comparison. San Francisco had endured a major crisis and suffered major losses. The city had the strong case it needed to justify a major investment that would reduce its vulnerability to similar events in the future. It would be hard to argue that improvements to their water supply system were not prudent and necessary.

Seven years later, in 1913, the US Congress passed the Raker Act authorizing San Francisco to make improvements to the Hetch Hetchy valley and Toulumne River to provide a water reservoir for the public good. These improvements took the form of the construction of the O'Shaunessey Dam on the Tuolumne River near the mouth of the valley. The dam was a monumental project in its day. Upon completion of the dam in 1923, concrete walls stood 227 feet above the valley floor and reached 113 feet belowground. At its base far underground, the width of the dam is equal to the length of a football field and stands on bedrock to provide the stable foundation necessary to resist the weight of hundreds of feet of water held back by its concrete form. This towering structure was eventually raised even higher to 312 feet in 1938. The curved walls of the dam now hold back over 360,000 acre-feet of water collected from a 459-square-mile watershed of melting snow within Yosemite National Park. These walls now constrain enough water to provide more than two million people in and around San Francisco with a majority of the water they need. These walls also created an excellent hydrologic reservoir for generating electric power. Power stations were constructed near the dam to harness the energy of the rushing water released by the upstream reservoir. The dam and its associated power plant now provide 500 megawatts of power to the California electricity grid.

Lost in the construction of this great dam was the natural organic beauty of the Hetch Hetchy valley floor and the ecosystem that thrived there. The stands of Black Oak trees and pleasant meadows that once flirted by the banks of the Tuolumne River were suddenly submerged by the oppressive weight of tons of water. The hikers' path and anglers' spot along the river were no more. The animals that once thrived in this valley were forced to relocate or drown.

The preservationists' defeat at Hetch Hetchy must have been a shock to the movement and widened the rift between the competing ideologies. The loss of Hetch Hetchy in 1913 was soon followed by the loss of the river valley's famous protagonist, John Muir, who died in 1914. The defeat also underscored the general view within Congress and society at large that preserving nature was fine to an extent as long as it did not compromise other human needs. The defeat showcased the youth and shallowness of the preservationist movement and how preserved national assets were not truly out of reach of competing interests. The rest of the country simply did not share the preservationist agenda much beyond wanting a nice place to vacation.

The conservationists' victory was probably not a true endorsement of conservationist ideology. Instead, it represented a preference of conservationism over preservationism when the two came into stark conflict. It is difficult to imagine Congress suddenly adopting a conservationist ethic when called to vote in 1913. It is easy to imagine them valuing the welfare of the inhabitants of a large West Coast city over the aesthetic quality of a small mountain valley that most of the members of Congress had probably never seen. It is worthwhile to remember that, in the early 1900s, most of the population of the United States lived east of the Mississippi River (US Census Bureau 1911). Most of the members of Congress naturally came from the eastern half of the country as well. For all of our relatively short national history, we had primarily been an East Coast society. The East Coast contained our seat of government and our largest, most well-known cities. While our nation was founded on ideals of political independence and citizen representation, it soon became a prosperous nation valuing economic and industrial development. The common goal was to build a wealthy society, not to preserve, or even necessarily to conserve, the natural resources found within our borders.

Hetch Hetchy was an early foreshadowing of the environmental movement to come a half century later. The loss of a pristine California mountain valley to the benefit of a reliable municipal water and electric supply for a great western city merely highlighted the environmental choices that would eventually come to the forefront of a new environmental discourse. Between this signature environmental conflict of the early 1900s and the emergence of the modern environmental movement in the 1960s, there were many social and political issues that commanded our attention. Upon reflection, it should come as no surprise that this infant environmentalism would fade from view for several decades.

The first half of the twentieth century was an extremely tumultuous time in America and around the world. Great wars in Europe, North Africa, and Asia stole millions of lives. World War I and World War II pitted the powerful nations of the world against one another. From 1914 to 1918 and again from 1939 to 1945, the European continent was engulfed in bloody conflict. Muddy trenches, barbed wire, poison gas, and insanely ineffective frontal assaults pitting human flesh against steel bullets spitting from rapid-fire machine guns were the hallmarks of the first Great War. The second Great War saw the development of more advanced weaponry with tanks, airplanes, and even the advent of nuclear weapons. Neither Great War was confined solely to Europe but spread beyond to North Africa, Asia, and the Pacific. Epic battles in both great wars led to an immense death toll and the widespread destruction of cities, towns, factories, and farms. Nazism flourished, and millions of lives were cruelly exterminated in an ill-fated attempt to eliminate selected groups of innocent civilians. Nuclear bombs erupted over two Japanese cities, killing thousands in mere seconds and introducing humanity to a terrifyingly destructive power that most had never conceived of before—and a power that still haunts us today. Figure 2.2 shows the distinctive form of a mushroom cloud. This term probably had little meaning to most people before Nagasaki and Hiroshima but has now come to be associated with nuclear bombs. Is it any wonder that conservation, preservation, and environmentalism did not register on our radar screens?

The US stock market collapse of 1929 and the subsequent Great Depression lasting into the late 1930s stole millions of futures. The US economy faltered, and our sense of security and worth was jarred to the core. People lost their jobs in droves, families lost their homes, and our society changed. Suddenly, there were lines of people in front of banks, at the doors of unemployment offices, and waiting for a meal at soup kitchens across the country. We went from being carefree to being careful, very careful, when deciding how to spend the money we had, for who really knew how long it might last. Life's luxuries became unobtainable for many. Even

FIGURE 2.2 Artistic rendering of the mushroom cloud resulting from a nuclear explosion. (Illustrated by Mark Benesh. With permission.)

life's necessities seemed to slip out of reach. We had to make do with what we had and hope that it would be enough. Is it any wonder that the environment did not make the list of life's necessities?

In the 1950s, McCarthyism and the communist threat stole some of our humanity. For a few insane years the US Capitol was distracted and disturbed over the perceived threat of communist activity within the nation. In a series of congressional witch hunts, personal liberties and reputations were sacrificed in the name of this ghostly threat. We Americans grew fearful of the Russians and the threat of nuclear war. We built bombs and bomb shelters. We stockpiled food and learned how to avoid radiation. We sent soldiers halfway around the world to Korea and Vietnam with proclaimed intent to stop the communist advance. America went 0-1-1 in those bloody conflicts; zero wins, one loss, and one tie. Is it any wonder that environmentalism simply did not measure up?

These extremely traumatic sociopolitical events captured an enormous amount of public attention during the first half of the twentieth century. There was little time or intellectual energy left over for anything else, much less for something so abstract as the environment that few had ever given much thought to before. The health of the environment simply did not compare to sons and husbands dying on foreign battlefields. It did not compare to rationing of many of the normal goods and services that we used in our everyday lives. It did not compare to women and girls thrust into the national workforce to operate the factories and industries needed to feed the war machines. It did not compare to the loss of a job or a lifetime's savings being wiped away in a few hours' time. It did not compare to the thought of a communist invasion or the incessant worry of just when the nuclear bombs would start falling and were there enough supplies in the bomb shelter to ride out the first wave of deadly radiation. It just did not compare.

The twentieth century that had started relatively well for the environmentalists was quickly derailed by events quite beyond their control. The events of the first half of the century were so large and threatening that they were quite beyond anyone's control. Environmentalism took a forced hiatus, a 50-year vacation from the public's attention.

That began to change, however, in the 1960s, and what an appropriate decade to accommodate a change in thinking! Politically, John F. Kennedy and Lyndon B. Johnson were pushing the envelope of our society's normal ways of doing things. The Cuban missile crisis had been won, and a small measure of justice began to creep into our politics of voting and equality. We had grown wealthy and powerful as a nation and could afford to reflect a little about the nature of our world and our place within it. We had become embroiled in war in Vietnam, but it was a very unpopular and unsatisfactory war. The rhetoric that this war was to defeat the communist menace began to grow thin and unravel. This war did little to rally us around our government. In fact it encouraged the opposite, and generated popular protest on college campuses, cities, and towns across the nation. America seemed poised and ready for change, and amid all these powerful and emotional events, a river caught fire.

The Cuyahoga River in northeast Ohio near Cleveland caught fire on June 22, 1969, and caught the attention of *Time* magazine (*Time* 1969). It would have caught my attention, too, if I had been the least bit aware of the environment in 1969, but I was 6 years old then and had other things on my mind. Had I been told about the event, even discounting the fact that I was only 6, it would not have made any sense at all. Everybody knows that rivers do not burn, or at least should not burn. Water is what we use to put out fires, after all, and I am pretty sure that I understood that concept at 6 years of age.

Of course, it was not the water itself that caught fire but flammable materials in and on the water that caught fire. Industrial waste products discharged into the river caught fire. This single-day event was probably not as much an environmental catastrophe as it was an environmental awakening. As a society we finally stopped hitting the snooze button on the environmental alarm clock. We finally roused ourselves from a peaceful state of unconsciousness and opened our awareness to things that should not be. A river should not catch fire. Perhaps these industrial wastes should not be in the river at all.

In fairness to history, this was not the first time the Cuyahoga had caught fire. Nor was it the costliest or deadliest. Reportedly, the picture run in the *Time* article was actually of a fire in 1952 that burned for 3 days. The 1969 fire burned only 30 minutes, not much time to get a good photograph in the days before cell phone cameras (Nordhaus 2007).

It turns out that the *Time* magazine article caught the attention of America and our political institutions. Environmental laws, regulations, and governmental oversight that had begun to come into being during the 1960s accelerated sharply in number, scope, and importance during the beginning of the 1970s. The National Environmental Policy Act was enacted in 1969, and the United States Environmental Protection Agency (USEPA) was formed in 1970. The Clean Air Act of 1970 greatly strengthened federal and state governments' authority to protect air quality—protection that was sorely lacking in previous versions of the legislation in 1955, 1963, and 1967. The federal Insecticide, Fungicide, and Rodenticide Act, the Clean Water Act, and the Noise Control Act were instituted during 1972. The Endangered Species Act came into being in 1973, and the Safe Drinking Water Act joined it in 1974.

Perhaps the most influential of these government acts was the National Environmental Policy Act (NEPA). Its importance stands in stark contrast to the number of Americans who know of its existence. NEPA mandated two relatively simple things. First, it required all federal projects to complete an environmental assessment of the probable environmental impacts of the project, and second, it established the Council on Environmental Quality that reports directly to the president of the United States (Council on Environmental Quality 1970). While the Council had the president's ear on environmental matters and undoubtedly raised environmental awareness within the executive branch of the US government, the requirement for detailed written environmental assessments has had the greatest lasting impact. This requirement has been applied generally to all projects that receive federal funding. Title I, Section 102, Part C is the heart and teeth of the NEPA. This section spells out the requirement for a detailed written report that includes

1. The environmental impact of the proposed action
2. Identification of any unavoidable adverse environmental effects if the proposed project is carried out
3. Alternatives to the proposed action

4. An analysis of the short-term impacts to the environment as compared with the long-term productivity of the environment

5. Identification of any irreversible and irretrievable usage of resources involved with the proposed project

These reports are typically referred to as environmental impact statements. not only did these environmental impact statements improve our level of analysis of proposed projects that had potential environmental impacts but the requirement for their preparation set a legal standard for environmental analysis that was subject to court review. Federal projects, or any project that received federal funding, could now be held accountable for how completely they performed this assessment. Environmental interest groups could now challenge local, state, and federal agencies and departments over their environmental impact statements. They could take these agencies to court arguing that their environmental impact statement was incomplete, and petition that the proposed project be put on hold until a proper environmental impact statement was done.

Note that the NEPA did not require that any certain finding be included within the environmental impact statement. It did not require that the project be "good" for the environment in order to proceed. It simply required that a comprehensive analysis regarding environmental impact be done. While this may not seem good enough from an environmentalist's point of view, it has given environmental advocacy groups another weapon in their arsenal for resisting environmental degradation. Court action can stop a project in its tracks, and the NEPA opens up more possibilities for court action. An environmental advocacy group may wish to slow or stall a project in hopes of having more time to lobby politicians and gain support from their followers. The NEPA also forces an environmental impact statement to be written that, in itself, can spur the debate about the critical environmental issues related to the proposed project. NEPA has done much to highlight environmental issues over the past 40 years and to inform and educate us on the environmental impacts that are tied to our many forms of economic development.

Undoubtedly, these major governmental actions were not the sole result of a river catching fire. Rather, the Cuyahoga burning serves as an appropriate symbol of a turning point in the American way of thinking about the environment. The late 1960s and early 1970s witnessed this shift in attitude as more and more people became interested in environmental causes, and numerous environmental advocacy groups came into being. The following is a partial list of some of these action/advocacy groups that were established during this time, along with the year of their formation.

- Citizens for a Better Environment (1971)
- The Cousteau Society (1973)
- Earthwatch (1971)
- Environmental Defense Fund (1967)
- Environmental Law Institute (1969)
- Friends of the Earth (1969)
- Greenpeace (1971)
- League of Conservation Voters (1970)

- Natural Resources Defense Council (1970)
- Population Institute (1969)
- Union of Concerned Scientists (1969)
- Worldwatch Institute (1975)

Groups such as these offered individuals a way to become involved, or at least feel involved, in defending our world. Both the US government and a growing number of citizens deemed the environment worthy of protection. The government certainly had shown that it intended to be at the center of the effort, but the populace clearly was not going to trust the environment to the government without some help or at least informed oversight.

Grassroots environmental protection organizations have had a great influence on our current view of the environment. They have represented a myriad of concerns from acid rain to whales and from forest fires to an ozone hole above Antarctica. The smorgasbord of issues on their table invited us all to be involved with at least one issue that struck a chord within our hearts. And once we have personally identified with one issue, it is a relatively small step to care about another, and then another. These advocacy groups have moved from being pesky, left-wing fringe alarmists in the '70s and '80s to mainstream stakeholders in our current environmental debates. What began as grassroots activism has evolved into private, not-for-profit corporations that prosper on the charity of their donors.

While these groups have won some battles over the years, their real impact has been in gradually changing individuals' attitudes over the decades. The many protests, marches, sit-ins, and letter-writing campaigns produced either victories or losses of the issue at hand, but always increased our collective awareness of our environment. Gradually, this awareness has seeped into the souls of the people of the new millennium, and today we think about the environment in a way that is significantly different from how the people of the 1970s thought about it. Even those of us who lived during the '60s and '70s probably view the environment differently from what we used to.

This should come as no great shock to us, as it is quite natural for our cultural attitudes, views, and even morals to shift with the passage of time. What is interesting and significant is the way this cultural shift has occurred. Almost unconsciously we have accepted a radically new view of the environment. We have been so busy and so preoccupied with all the skirmishes of the day, acid rain, damming wild rivers, logging in the national parks, ozone holes, and so forth, that we failed to consciously acknowledge that we were tacitly accepting the premise that the environment was at risk and that we needed to fix it. The war for environmental relevance and responsibility was steadily being won. With every newscast showcasing the battle, with every student entering college, and with every soul born into our world, the balance of thought has tipped more and more toward environmental protection.

This has been a messy transition. Political administrations have landed on one side or the other of the environmental spectrum. Laws have been passed or repealed, environmental programs have been funded or ignored, and debate has raged over the "proper" amount of environmental oversight. Sometimes environmentalists are happy, sometimes capitalists are pleased. On the whole, environmentalism has marched steadily onward, sometimes quickly and sometimes more slowly, but always eventually trumping the political posturing of administrations that try to put the brakes on environmental progress.

Not only has environmental awareness and consciousness become mainstream but the environmental issues themselves have grown and expanded in scope and importance. In the past, we focused almost exclusively on local and regional problems; now global issues occupy center stage. We used to be primarily concerned about air pollution above our cities and towns. Now we are told of air pollution generated half a world away and carried around the world in the upper atmosphere to eventually be deposited over nations near and far. We used to be concerned about declining water quality in our lakes and streams. Now we know about oil spills in foreign seas and controversial harvesting of whales far from our shores. We now know that we can manipulate the genetic makeup of the crops we produce and the livestock we raise, and nations haggle with one another over the quality and safety of food that is imported into their markets. We are now aware of ozone holes above the poles, and the issue of global warming—or global climate change as it has become known— has become a very obvious example of the broadening range of environmentalism.

The new millennia saw a general human acceptance of the environmental calamities presented in the previous decades. Debate over cause, effect, and responsibility generally vanished from the discussion. It was enough to know that humanity was responsible for the state of the environment and that something must be done to avert the pending disasters.

There are a few major lessons to be learned from the last few decades of the twentieth century. The first lesson is that the central government has taken the lead role in protecting the environment. Many would argue that the government has been very effective in this regard. The question that usually arises however is, Has the government done enough? That is a fair question that often leads to a debate over one particular issue or another. The debate is part of the process of environmental decision making. I hope the four pillars that we will discuss in the rest of this book can help inform that debate.

The second lesson is in the emergence and importance of environmental activist or environmental advocacy groups. They have publicized a long series of environmental issues and broadened the debate to transcend governmental institutions and include a large number of individuals that historically had little voice in these matters. These activist groups created a way for people to become involved and exert some influence far beyond the normal bounds of a single person. Environmental advocacy organizations have increased in number, scope, relevance, and acceptance. They have become mainstream organizations in our societies. They are no longer defined by confrontation, but instead often interact with government and industry in a rather civilized consultative manner. Their acceptance mirrors the general acceptance of environmentalism within our societies.

The third lesson resides in the general shift in human values and ethics regarding the natural world. This has been an amazing transformation. In just a few decades we have moved from a state of ignorance, even oblivion, to a belief that humanity is largely responsible for the scars we have inflicted upon the earth. This environmental movement is not fully embraced, but it sure seems close. Being fully ingrained in our culture is far different from being fully embraced by our people, but it has been accepted well enough to play a prominent role in many aspects of our lives. Perhaps it has come down to a question of simple economics. In general, we all prefer a healthy and clean environment. The real question is how much does it cost, and are we willing and able to pay for it.

It will be quite interesting to see what the future holds for environmentalism. Will the much publicized disasters come to pass, or will our environment merely evolve slowly and steadily as humanity manages the challenges that present themselves? It is extremely difficult to imagine our environment and our relationship to it remaining stagnant. Think how much things have changed in the past 100 years. What person living in the year 1900 could have predicted cloned animals and enough energy contained within a single bomb to annihilate the inhabitants of a city? What person living in the year 1950 could have predicted cell phones, the Internet, and space travel? What will we be surprised to find when we ask similar questions of people living in the year 2000?

Now that we have touched upon the beginnings and progression of our modern environmentalism, the next chapter will begin to discuss the ethics behind it. We have learned a little bit about where environmentalism came from. We still need to examine what is behind it. What within us ultimately supports our views of the environment and allows us to act in defense of the world we share?

REFERENCES

Marsh, G.P. 1864. *Man and Nature or Physical Geography as Modified by Human Action*. New York: Charles Scribner.

Williams, G.W. 2005. *The USDA Forest Service—The First Century*. Washington, DC: USDA Forest Service Office of Communications.

National Park Service. 2005. National Park System Areas Listed in Chronological Order of Date Authorized under DOI.

National Park Service, US Department of the Interior. 2006. Federal Historic Preservation Laws: Antiquities Act of 1906 as Amended.

Righter, R.W. 2010. *The Battle over Hetch Hetchy: America's Most Controversial Dam and the Birth of Modern Environmentalism*. New York: Oxford University Press.

The Sierra Club. 2010. Hetch Hetchy Timeline. http://www.sierraclub.org/ca/hetchhetchy/timeline.asp (accessed September 6, 2010).

Hansen, G. and Condon, E. 1989. *Denial of Disaster*. San Francisco: Cameron & Co.

US Department of Commerce, National Oceanic and Atmospheric Administration. 1972. A Study of Earthquake Losses in the San Francisco Bay Area—Data and Analysis, A report prepared for the Office of Emergency Preparedness.

US Census Bureau. 1911. Area, Natural Resources and Population. http://www2.census.gov/prod2/statcomp/documents/1911-02.pdf (accessed September 6, 2010).

Time. 1969. The Cities: The Price of Optimism. http://www.time.com/time/magazine/article/0,9171,901182-1,00.html (accessed September 6, 2010).

Nordhaus, T. and Shellenberer, M. 2007. *Break Through: From the Death of Environmentalism to the Politics of Possibility*. Boston: Houghton Mifflin. pp. 22–24.

Council on Environmental Quality. 1970. The National Environmental Policy Act of 1969. http://ceq.hss.doe.gov/nepa/regs/nepa/nepaeqia.htm (accessed September 6, 2010).

3 The Ethics of Environmentalism

Environmentalism today exists in many forms. Consider the extreme tactics of groups such as EarthFirst! versus the comparatively benign approach of The Nature Conservancy. One has historically been associated with acts of sabotage and violence while the other promotes the purchase and accumulation of critical areas of land. Certainly these two organizations must operate under a different code of ethics, a different set of morals, and a different view of right and wrong.

The ethics of environmentalism is a very broad term. It tries to account for the forces within us that drive us to think, feel, and view the environment in the ways that we do. It recognizes that there are many ways in which people practice environmentalism and therefore there can be many ethics as well. Some environmentalists focus on regulation and strict limits to pollution while others offer free-market solutions to curb emissions. Some focus on technology while others have a strong belief in natural forces and natural solutions. Some are content to focus solely on one issue while others fully intend to save the world. Some are very thoughtful in examining their feelings and beliefs while others are content to react to the environmental supplications that are thrust upon them. Today's range of environmental ethics is certainly wide and at times quite dissimilar.

The term itself, "ethics," with its basic definition of delineating right from wrong, is typically too neat and tidy for environmental issues. Environmental ethics is an oversimplification yet a very convenient one. It allows us to describe some of the basic values, views, and beliefs that we hold with respect to our relationship with the natural world. It also forces us to defend the logic and rationale of our approach when we begin to act on behalf of the environment.

This book is about a specific type of environmentalism—practical environmentalism. Its ethic is deeply rooted in the independence of the thoughtful individual. Its ethic is a broad one. It is not solely confined to the examination of environmental damage but rather considers environmental degradation on an equal footing with its other three pillars of resource conservation, economic progress, and personal benefit. It is comfortably incorporated within the aforementioned greenish shades of gray.

However, before we enter into a more comprehensive examination of the ethic of practical environmentalism, perhaps it is best to begin with a brief accounting and discussion of the wide variety of environmental ethos that exist. In an attempt to simplify the broad and varied landscape of environmental approaches, I will present just a few broad categories for discussion that hopefully will encompass the majority of ethics that exist. These categories are

1. Issue-driven environmentalism
2. Process-based environmentalism
3. Perspective-based environmentalism

We will then discuss the ethics of Practical Environmentalism and see how it fits among these broad categories. As we go through these categories, it may be helpful to keep a number of basic questions in mind that we could ask as we consider the ethics of any particular brand of environmentalism.

- Are we masters of the natural world or members?
- Can we control nature and its myriad functions, processes, reactions, and phenomena, or are we merely participants in a grand game that we do not really understand?
- Are we responsible for the other species that share our planet?
- Are we responsible to the generations that follow us?
- Is there some sense of environmental fairness or justice that should include compensation for our differences in wealth, technology, and history?

These questions, among many others, are appropriate considerations when defining and justifying just how we will interact with our environment. These types of questions could be helpful in discerning the differences between particular forms of environmentalism. There really are many forms of environmentalism. Of these many forms, some are broad and inclusive while others are narrow and specific. Some seek to stand on a firm theoretical foundation while others choose to seek out the emotional parts of our beings and lay claim to our hearts. Let us begin our discussion by examining the ethics of issue-driven environmentalism, and, in much the same fashion as we began our study of the history of environmentalism, let us first talk about conservation and preservation.

ISSUE-DRIVEN ENVIRONMENTALISM

Conservation and preservation are early forms of issue-driven environmentalism. Both focus on natural resources, the second of our four pillars. Both are historically significant and are predecessors of our modern environmental movement. They called attention to the natural world and defended its importance to a society that largely took it for granted. Both also stressed the rather modern notion that humankind could damage these resources to a significant extent such that its capacity to provide benefit to humanity could be drastically reduced.

The most striking difference between these two early forms of environmentalism is their view of the benefits that nature provides. Conservationists valued nature for the economic resources it provided to humanity. To them, the real question was the continued abundance of resources necessary for human economic progress. Resources such as fertile soil, timber, and fresh water dare not be wasted, nor should the environment that provides these renewable goods be compromised for fear that the quantity of these goods be devalued as well. The conservation ethic extends to nonrenewable resources as well. These resources should be used at maximum efficiency to extend their inventory as long as possible. The ethics of conservation include human dominion over and responsibility for the natural world. Environmental degradation is important but primarily due to its impact on nature's capability to provide real goods and services for human consumption. The conservation ethic is one of utility. Nature is valuable and important primarily because of its usefulness to humanity.

Preservationists valued nature for significantly different reasons. To them, the greater value is the intrinsic beauty and wholeness of nature left in its undisturbed state. The coal should be left in the ground underneath the most beautiful western prairies. The magnificent redwood trees should be allowed to stand guard for all eternity up and down the foggy northern California coast. The most beautiful, natural, and wild spots of the globe should not be despoiled by destructive human actions in the name of economic development. Wildlife should have space and sanctuary to live, breed, and thrive. Wild things and wild places benefit humanity. Wilderness has a unique ability to restore the soul and rejuvenate the spirit. Undisturbed landscapes are only grossly undisturbed, however. These beautiful wild landscapes are to be saved for the hiker, camper, and tourist.

The central ethic of preservation is certainly nature focused. While humanity is very much the beneficiary of the natural state, the underlying ethic leans toward the value in nature itself more than human enjoyment of nature. The preservation ethic is generally distrustful of humanity's capacity to value nature appropriately and would prefer that there are firm and inviolate rules in place to protect nature from humanity's harm.

As noted in Chapter 2, conservation versus preservation existed as a healthy debate at the end of the nineteenth century and the beginning of the twentieth century. Ethical components of each can still be found in our modern environmentalism. Our modern form, as evidenced through environmental movements of the late twentieth century, has a much broader ethical range than the conservation–preservation dichotomy that preceded it. There are many environmental issues that cause concern today, and many of these have grown to be fairly well known.

Issue-driven environmentalism focuses on specific environmental issues, points out the problems, offers solutions, and demands changes. It is defined by the issue itself. This is the type of environmentalism that we hear most about. A species is near extinction, or the oceans are becoming more polluted, or our world is getting warmer. Issue-centered ethics revolve around particular environmental issues that have gained attention and notoriety, typically as a result of some group or organization publicizing the issue as being important. Indeed, these issue-centered ethics are often associated with particular environmental groups, often private groups that have built a loyal following of supporters who fund their activities. These are environmental action groups or nongovernmental environmental groups, sometimes referred to simply as NGOs, or nongovernmental organizations. These are groups with familiar names like The Sierra Club, or People for the Ethical Treatment of Animals (PETA), or Greenpeace. Many of these groups were founded as a response to a particular environmental issue.

There are many very interesting issue-centered ethics within the range of environmentalism today, many more than we can reasonably discuss here. Each issue and each group associated with the issue will have its own ethic. These ethics will reflect the membership, goals, and history of the group. Just as there is a wide variety of environmental issues, there is a considerable range of ethics. Issues are portrayed with differing degrees of criticality and importance. Environmental activist groups promote a variety of solutions and go to varying lengths to get their solutions noticed and accepted. What is generally common among issue-driven environmental groups is their reason for being in the first place. Specifically, the ethics of issue-driven environmentalism have typically been borne out of protest. We will discuss this more completely later in this chapter as we will include the ethic of protest environmentalism under the process-based category. Suffice it to say for the moment that the ethics of issue-driven environmentalism are many, varied, and defined by the issue and organization themselves.

PROCESS-BASED ENVIRONMENTALISM

Contrary to issue-driven environmentalism, process-based environmentalism is defined by the general approach to solving environmental problems. Where issue-driven environmentalism identifies and publicizes particular environmental crises, process-based environmentalism focuses on how we respond to the issue. It can be widely employed to address practically any environmental issue that we encounter. It works hand in hand with issue-driven environmentalism in that process-based environmentalism is focused on the method and mechanics of analyzing environmental issues and instituting change for the sake of the environment. There are two well-established process-based environmental ethics in our societies today. They are regulatory environmentalism and protest environmentalism.

REGULATORY ENVIRONMENTALISM

The regulatory ethic is a strong force within modern environmentalism and is widespread around the globe. The basic premise of the regulatory ethic is that individuals have no personal incentive to act appropriately with regard to the environment. It is individually inefficient to reduce the negative impacts of one's own activities when most likely those negative impacts will not be felt by the individual. The wood stove that keeps me warm at night emits some smoke into the air that drifts away from my home to slowly dissipate over my neighbors' properties. Economists would term this a negative externality, a negative result that is felt so slightly by the individual that there is no reason to behave differently. Therefore, it is the responsibility of the government to use regulations to limit pollution and preserve the environment. These regulations can take many forms, from banning certain activities, such as the use of the insecticide DDT, to limiting the amount or rate of pollution from other activities such as constraining particulate emissions from a factory's smokestack.

Within the regulatory ethic there exists an important subethic that incorporates some free-market capitalistic ideals within the realm of regulating pollution. This subethic suggests that governments should not act as the sole authority in deciding how to implement regulations to limit pollution or other negative environmental impacts. Instead, the government should set standards and limits that apply to a general geographic area or a particular industry and then allow those affected to determine the best way to meet these limits. This is a hybrid approach where market principles are allowed and encouraged to function within the overarching environmental regulatory framework. Cap and trade is a frequent synonym for this approach in the United States and has been applied to the regulation of sulfur dioxide (SO_2) and nitrogen oxide (NO_x) emissions from electric power plants in the eastern United States. In this case the US government regulates the total amount of emissions allowed, but individual emission sources still have some flexibility in determining how much they can emit. We will discuss this particular case in much more detail in Chapter 11.

The cap-and-trade logic is that while free-market forces are not sufficient to produce the desired environmental effect in and of themselves, these same free-market forces can be very effective and efficient in accomplishing the desired environmental effect once the boundaries have been established. In this subethic, the government's role is to establish the rules of the game and draw the boundaries around the playing field. Perhaps the government must act as a referee as well, but they do not tell the players what to do or how to do it. "Let the players play," say the free-market proponents. Let the players decide how best to satisfy the requirements with optimum economic efficiency. The government does not need to be too concerned with the details as long as the overall goals are met.

The free-market ethic includes a basic trust of industry and private corporations. The ethic approves the compromise that forces these entities to conform to the environmentally protective regulation while allowing them some freedom and flexibility in doing so. It includes the explicit recognition that economic considerations are important and that individual industries and corporations are best suited to managing the economic–environmental dilemma within the established regulatory bounds.

The purists of the regulatory environmental ethic would disapprove of the free-market amendment. The regulatory ethic in its simplest form asserts that the government is responsible for limiting environmental damage, or impact, at the source of pollution with the establishment and enforcement of strict limits. This ethic includes a general distrust of industry and private corporations. It includes a belief that regulated entities will only do the bare minimum to comply with the rules forced upon them and that some will lie, cheat, and steal to circumvent appropriate environmental controls in an effort to maximize their own profits and economic well-being.

A basic premise of using regulation to protect the environment is that it is necessary at times to force individual entities to comply with an action that is against their individual will. In this case, the benefit to the group outweighs the detriment to the individual. This is a common moral question in many areas outside of environmentalism, and an evaluation of appropriate representation in the political process becomes paramount in ascertaining whether this forced submission is defensible. Is the individual adequately represented within the political power structure such that his or her voice is heard? Is the decision to limit rights and freedoms reasonably made with appropriate debate and discussion? Or, do the preferences of select powerful minority groups outweigh all other considerations and drive egregious forms of command- and control-governmental activities?

A particularly contentious example of this coercion component of the regulatory ethic is the case of population control. Population control falls easily within the environmental arena following the logic that too many people exert too many demands on the resources of the natural world, which leads to the degradation and even destruction of the environment. Under this theory, it is preferable to limit the number of births within a population to avoid excessive growth and excessive demands upon nature. The means to limit births could easily fall within a regulatory framework and could range from economic sanctions to voluntary or involuntary medical procedures. In this example, it is easy to imagine how coercive power could be applied to the detriment of individuals.

The regulatory ethic exists at various scales and is present within local, regional, and national governments. Cities and towns enact environmental legislation. The small town where I live prohibits burning the leaves that fall from our many old trees. This local legislation is generally supported in my town, and the residents value the cleaner smoke-free air that results. Environmental legislation is most typically found at the state, provincial, and national levels. There are myriad laws, rules, and regulations that deal with air quality, water quality, habitat preservation, and construction permitting at these levels of government. Regulation even exists within confederations of national governments and has been used in multinational agreements to deal with global environmental issues such as the control of ozone-depleting substances in the Montreal Protocol of 1987 and agreements to reduce greenhouse gases emissions in the Kyoto Protocol of 1997.

The regulatory environmental ethic is certainly common and pervasive at the present time. It is a rather general ethic in that it deals mainly with implementing rules to achieve desired outcomes. It does not explicitly take sides on particular environmental issues but instead provides a forum for debate, then undertakes government action to achieve the hoped for results. While the regulatory ethic operates in the human political arena, it is not really human centered. It does not focus on the individual but rather seeks to implement rules and regulations common across towns, cities, states, nations, and even worldwide organizations. It discounts individual values and preferences in the name of conformity and fairness. It merely strives to provide environmental regulation that is politically palatable.

PROTEST ENVIRONMENTALISM

Protest environmentalism is the other main process-based ethic. It exists as an emotional response to perceived injustice toward the environment, or a perceived incompetence and ineptitude of groups that impact the environment. Protest environmentalism rails against the corporations that despoil nature and the governments that permit it. Protest environmentalism exists within a very broad range, from peaceful letter-writing campaigns to acts of vandalism or confrontation. At its extreme, it drives people to take to the seas to track, follow, and harass whaling ships. It leads people to break into research and development laboratories in the dead of night to free the test animals used there. It convinces people that it is somehow just to drive large metal spikes into the trunks of magnificent trees so that loggers and mill workers will fear the explosive result if their saws should come in contact with the metal hidden within the wood (ultimately in the hope that the trees will not be logged in the first place).

In its more benign form, it encourages people to raise their voices through demonstration and communication to convince others that the perceived injustice should be remedied. We can call or write to our elected officials. We can carry picket signs and hope for a few seconds of TV time on the nightly news. We can enlist the help of celebrities to carry the banner for our cause.

Protest environmentalism thrives on the organization and association of like-minded individuals. Again, the acronym NGO, nongovernmental organization, is a useful abbreviation to describe these groups. NGOs form through the energy and enthusiasm of a few and can grow to thousands of members. Members often participate directly in the actions of the organization, but many often simply offer monetary contributions in support of the group's purpose. Protest environmentalism via NGOs has capitalized on our individual willingness to show support for a cause. It has allowed us to support the group's activities without demanding very much from us. We can participate in the environmental protest from the comfort of our favorite chair. Protest environmentalism has shown the ability to tap into a large, if somewhat silent, segment of society with an interest in the environment. Large nationwide memberships of some NGOs have strengthened these organizations and increased their standing and power within our social and political structures.

Protest environmentalism thrives on attention. It seeks to convince others that change is needed. Attention and publicity aid the cause. Its underlying ethic is one of conversion. If only people were aware of what was happening and understood the ramifications, they would join the group and help convince even more people that change is needed. As more and more people join the cause, it becomes easier and easier to influence society to change. Perhaps governments will feel sufficient pressure to enact new laws and regulations. Perhaps corporations will sense the change within society and change themselves in an effort to better market their products and services. Perhaps enough people will begin to think in new and different ways, and the perceived environmental injustice will fade into history just as some social injustices like slavery or genocide have begun their shameful retreat into the far reaches of our memory.

In some cases, protest environmentalism can border on coercion. If convincing others does not produce the desired results, perhaps we should force others to go along, either through peer pressure or through the force of law. Our tactics may become more about the ends than the means. Why waste time and energy convincing a "moral" majority if we can take a shortcut to get people to behave the way we want? Governments and the force of law and regulation are the typical allies when the ethics of protest environmentalism include a degree of coercion. Let us convince just a relatively few powerful government officials of the justness of our cause and get them to institute the laws necessary to make everyone conform to our vision of what the environment should be.

Like regulatory environmentalism, protest environmentalism is a process-centered ethic. However, where regulatory environmentalism is often a response to environmental issues, protest environmentalism is usually founded on the issues themselves. As mentioned previously, groups form because of a perceived injustice. So, at its core, protest environmentalism rallies around environmental issues and at times may appear to be issue centered. This complexity in categorization occurs because protest environmentalism has become so wide ranging that it incorporates many, many issues. To claim that protest environmentalism is issue centered is a vast oversimplification because there are far too many issues to fit within one nice, neat title. It is not very difficult to find NGOs that have competing objectives, which makes it difficult to lump them together under an issue-centered category. There are groups that fight for the societal survival of small family farms, and there are groups that feel that raising animals for slaughter is wrong. There are groups that advocate the importance of implementing renewable energy, and there are groups that are against despoiling the rural landscape with large windmills.

Protest environmentalism is not generally a human-centered ethic. The individual is not terribly important beyond what he or she could do to further the cause. The individual is expected to conform to the ideals of the group. At its core, protest environmentalism exists because environmental issues exist. The central theme revolves more around the issue itself as opposed to how the issue impacts humanity.

Protest environmentalism usually focuses on specific issues that are generally most applicable at broad scales. Certainly there are local groups protesting about local issues, but the ones we hear most about tend to be national or even international organizations trying to raise awareness and cause change at national and international levels. The power of the media and the scope and influence of government are greater at these levels.

So far we have focused on the ethics behind conservation and preservation, and we have touched upon the ethics behind environmentalism that is focused on the specific issues themselves. We have also discussed the process-centered ethics of regulation and protest. All of these ethical approaches, however, are founded and based on individuals' perspectives about the environment in general. So, let us now discuss a few perspective-based environmental ethics. These are the ones that often define the way we think and feel about the environment. These are the ethics that structure the reasoning behind our environmental actions.

PERSPECTIVE-BASED ENVIRONMENTALISM

Perspective-based environmentalism is the root of our system of environmental values. It is driven more by how we see and understand the world, instead of any particular environmental problem or potential solution. It is our lens for viewing nature and our place within the natural world. Understanding and analyzing specific forms of perspective-based environmentalism can often be facilitated by asking particular questions such as, does nature have "rights," does humanity have the capability to solve environmental problems, or how important is the environment when compared to other facets of our lives?

The list below shows a few of the concepts in the perspective-based environmental ethos that deserve mention and that will hopefully grant us an awareness and appreciation of the broad range of environmental views that exist today.

1. The Nonbeliever
2. The Live Earth
3. Spaceship Earth

4. Back to Nature
5. Doomsday
6. Anticonsumption
7. Cornucopia

THE NONBELIEVER

Forgive me for the title "nonbeliever" to label this perspective. It is certainly trite and admittedly an overgeneralization. I use the term to categorize those who either do not care very much about environmental issues or who feel helpless to do anything about them. These are the lost souls, the heathen that environmentalists try to enlighten and convert. It describes a state of mind where environmental thought, much less environmental action, never rises to the forefront. Nature and the world around us simply cannot compete for our attention with all the other normal priorities of our lives. Thoughts of our spouses, families, loved ones, jobs, social life, hobbies, and possessions do not have to make room within our hearts and minds for thoughts about the environment. So the nonbeliever perspective is really the antithesis of environmentalism. Figure 3.1 is included as a hopefully amusing label for this particular environmental perspective.

I really do not know how prevalent this perspective is within our societies. I imagine it varies from place to place and from generation to generation. I suppose we could conduct a survey and get a better sense of just how large this group is. I bet someone has already conducted such a survey. There might be many surveys that we could choose from. My guess is that this perspective began to diminish along with the "birth" of modern environmentalism in the late 1960s. This perspective is relatively easy to camouflage, though. It is easy to talk the environmental talk without walking the walk. It is easy to respond to a survey with the "correct" environmental answers while having no intention of acting in a way that benefits the environment.

The point here, however, is not to complain about those who do not share an interest in the environment. Rather my objective in discussing this perspective is to acknowledge its existence because it can make a difference in whether environmental action can occur and whether it can be successful. The nonbeliever perspective, through inaction, can influence the outcome of environmental programs and policies. It can deny funding. It can sway votes in the legislature. It can reduce momentum and deny issues the critical mass of energy and enthusiasm they need to move forward.

Enough of the negative, though. This is a book about environmental improvement, after all. Let us turn our attention to more positive perspective-based environmental ethics.

THE LIVE EARTH

"The Live Earth" approach to environmentalism considers the earth to be a living being. Note that this does not say the earth is filled with living things but rather the earth itself, in its totality, is a living thing. Phrases such as "Mother Earth" and "Mother Nature," so common in the English language, flow from this ethic. This ethic has roots in European as well as Native American history. It also appears in

FIGURE 3.1 The Nonbeliever. (Illustrated by Mark Benesh. With permission.)

modern times as the Gaia hypothesis and theory as proposed by James Lovelock (Lovelock 1975). For our discussion here, let us also use the phrase "Living Earth Theory" to describe the approach. Figure 3.2 shows a representation of this perspective. It was a hit with my children who immediately named this character "Globey."

This very earth-centered ethic values nature somewhat like a person. Nature itself deserves respect and consideration just like we would respect and consider another human being, or at least like we should respect another living being. Living Earth Environmentalism does contain a range of approaches from the spiritual to the scientific. Incorporating human traits of kindness or cruelty, generosity or injury, the more esoteric views within the Living Earth Theory grant nature the status of personhood

FIGURE 3.2 The Live Earth. (Illustrated by Mark Benesh. With permission.)

or greater. Nature can even become a magnanimous, forbearing deity granting us prosperity if we respectfully act within natural laws.

The more scientific end of the living earth spectrum often stops short of personhood or God status and instead focuses on the functionality of the natural world. For example, Lovelock asserted that the earth in its totality worked together to create conditions favorable for life. The rocks and soil beneath our feet, the oceans that surround the lands we inhabit, the sky above our heads, and the plants and creatures that share these places with us all work in conjunction with one another to create physical, biological, and chemical conditions that benefit and support one another. The Live Earth approach draws support from the circular nature of many of the earth's processes. For example, let us trace the hydrological cycle in very simple terms and follow water's journey.

We will begin with the great oceans. Most of the water on earth is stored here. It would generally remain here except for evaporation. The atmosphere that surrounds the earth allows sunshine to penetrate and fall on the earth's surface below. The atmosphere contains a number of gaseous elements that are relatively transparent to incoming solar radiation but are also relatively reflective to heat radiated from the surface of the earth. (This is the foundation of the global warming theory that we will discuss in later chapters.) This property of the atmosphere keeps the earth's surface warmer than it would otherwise be and therefore more hospitable to life as we know it. The solar radiation that strikes the oceans' surface evaporates water, which rises back up to the atmosphere where it is stored in clouds. From time to time clouds will release their stored water and precipitate it back to the surface in the form of rain, snow, sleet, ice, or hail. Once back on the surface, water will seep into the earth, flow to collecting streams and rivers, be soaked up by plants, or fall directly into lakes or oceans. Water that seeps into the earth will often find its way to an aquifer near the surface and may return to the surface through a spring far distant from where it fell. Water that collects in rivers and streams will eventually find its way back to the great oceans, and the cycle is complete and ready to begin anew.

In this example the oceans act primarily as a storage reservoir—one that also happens to contain a significant amount of animal and plant life. The atmosphere is the gatekeeper for allowing energy into the hydrologic system and serves to distribute water over the earth's surface as it transfers it from the oceans to continents. The landmass acts as a filter for water as it drains back to the oceans. The land, of course, is also home to many plants and animals, including ourselves. Its filtering properties allow some water to be stored while the excess drains away. Plants and animals then have access to the water without being inundated by it. This system of land, sea, and sky works together to store, transport, and distribute water in a manner that can be used by living things.

There are many other biological, chemical, and physical cycles we could examine beyond this simplified version of the hydrological cycle. The Live Earth view appreciates these cycles and the way they synergistically interact to provide for the inhabitants of the earth. This view maintains that we should study and better understand these cycles and align our activities to flow concurrently with them instead of fighting upstream against them. Whether or not we attach human characteristics to nature, we should at the very least appreciate nature's complex functions that support life itself. The Live Earth view would truly appreciate the 1970s television ads in the United States that featured the phrase "It's not nice to fool Mother Nature!"

Let us now turn our attention to another earth-focused environmental perspective. This one does not consider earth to be alive but rather considers it a vessel with limited capacity to support life.

SPACESHIP EARTH

Another interesting twist to our modern environmentalism is the Spaceship Earth view. This view has been dramatically rendered by earth photography from space. Space exploration via satellites and manned spacecraft permitted this unique perspective of our planet, a seemingly backward glance from an otherwise outwardly focus on what exists beyond our globe. Figure 3.3 tries to capture the essence of this view.

The Spaceship Earth view is a dramatic reminder that the earth really is finite. It took the modern space program to enable us to get far enough away from our own planet to notice how small it really was. In its entirety our planet fits neatly within a 5 × 7 photo. For many, these photos were an awakening, a near spiritual experience of how humanity is crowded onto a small orb. Discounting the space age promise of exporting humanity to other planets, we really have no place to run. We are stuck on this planet and bound by its resources. If we mess it up, there is no place to go, there is no safety net. If we fall, there will be no one to help us get up.

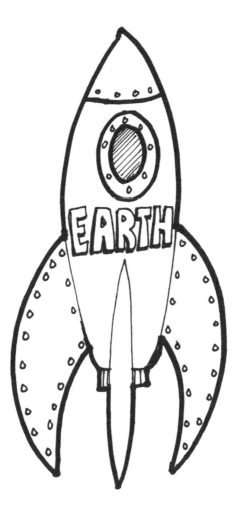

FIGURE 3.3 Spaceship Earth. (Illustrated by Mark Benesh. With permission.)

The Spaceship Earth view is quite similar to Lifeboat Ethics. That is, if you are stranded at sea in a lifeboat, you are quite aware of the limitations of your immediate surroundings. Food is precious, water much more so. Shelter from the sun's intense rays or the storm's wind-driven rain is critical for your survival. The very integrity of your vessel gives you a chance to survive. Without it, you would surely drown.

Lifeboat Ethics, often a study in psychology, can be extrapolated to the Spaceship Earth view. Now, the earth is our lifeboat. The integrity of the earth is our only chance for survival. The earth's resources become our food, our drink, and our shelter from a harsh existence. Before our ventures into space, our terrestrial vision and perspective would not easily allow us to make this association and this leap within our thinking. It took that retrospective photo from a spacecraft to make the association clear.

If we follow the logic of the Spaceship Earth view beyond the iconic photograph, we encounter an overriding theme of constraint and limitation. The harshness of the vast metaphoric sea around the lifeboat does not represent a beneficent, ever-giving nature but rather a future of desperation and want brought about by a growing human race with a growing appetite for nature's resources. Our excessive growth and desire eventually brings competition, depletion, and overuse of nature's abundance. Spaceship Earth is quite a nice lifeboat if we do not have too many people weighing it down and fighting over the meager rations stowed away. The Spaceship Earth view is ultimately a wake-up call to remind us that we really should take better care of our home planet. It is a symbol and implicit message all in one.

BACK TO NATURE

Another unique and interesting environmental perspective is the Back to Nature viewpoint. This particular view is somewhat reminiscent of the past in ascribing environmental benefit to the way humanity lived decades or centuries ago. This view finds fault with many of our modern social and economic norms. For example, millions of people now sit in their personal automobiles every day in a slow tedious commute to and from their place of employment. The food we eat is often produced on corporate, highly mechanized farms using large amounts of man-made fertilizers and pesticides. When we retreat to our homes in the evening, we shut ourselves up in a pool of conditioned air and stare at the television, or the computer, or our fancy new telephone that allows us to surf the Internet or check our email. Figure 3.4 symbolizes this view in an image of a farmer working the land the "old" way.

The Back to Nature view tends to have somewhat of an antitechnology bias. It often sees technology as a root cause of our environmental issues instead of as a potential savior. The development and evolution of the automobile have given us traffic jams, congestion, road rage, and tons of pollution emitting from the tailpipes of our many cars stuck in traffic. Our modern agriculture and its preference for large-scale mono-cropping production systems threaten the loss of natural biological diversity and spread tons of chemicals onto our land. Small family farms are becoming the exception, and large corporately owned farms the norm. All our electronic

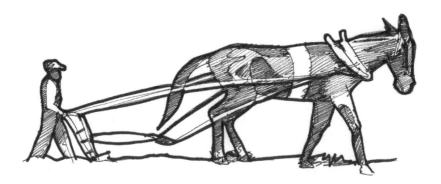

FIGURE 3.4   Back to Nature. (Illustrated by Mark Benesh. With permission.)

devices that we rely upon for comfort and entertainment create a demand for electric power that is at least partially satisfied by many coal-burning power plants with their tons of associated pollution entering our atmosphere. The front porch gatherings of friends and family on warm summer evenings have largely faded from our memories, replaced by the convenience of technology.

Back to Nature proponents tend to criticize many of our modern social and economic choices, making the point that our modern ways have led to environmental distress. In many ways, it would be difficult to argue with their logic. There are numerous examples of environmental distress that are linked to our technological "advances." Detractors of this view might argue that rewinding the clock would do little to alleviate environmental distress and may cause other more severe problems as we struggle to provide for the world's population.

Of course, this view, like many others, exists in moderate to extreme forms. For example, some might advocate that we consumers should buy locally grown organic food. This will reduce the pollution caused by transporting food from half a continent away and lessen the pollution caused by spreading fertilizers and pesticides on the land. Others might go a step or two further and suggest that we depopulate the cities and send people back to the land to grow their own food like our ancestors did. Both options fall within the range of the Back to Nature ideology. They both include the historical reference of a time when we ate what we grew, what our neighbors grew, or what came from nearby farming areas.

The Back to Nature view contains a viable tension around the ideas of technology and history. It is suspicious of modernity and humanity's capability to influence and control nature without causing undue stress to the environment. It is certainly more nature focused than human focused, as the Back to Nature view proposes that we humans should alter our lifestyles to be more in tune with nature. It contains an implicit appreciation of the worth of nature in providing sustenance for people. The traditional time-honored natural ways are seen as valuable for the relatively slight impact they have upon the environment when compared with modern industrial approaches. The Back to Nature view does not explicitly include the Live Earth view, but it is certainly easy to see how these two perspectives could be seen as complementary.

DOOMSDAY

The Doomsday perspective has become more prevalent as environmentalism has moved into the twenty-first century. Its view is that catastrophe is just around the corner. It has found a home within the global climate change issue that has emerged to occupy the center stage of the modern environmental movement. The Doomsday view is a very emotional one, and one that gets a lot of press and publicity. It is sensational, and it grabs our attention. The Grim Reaper and the death of the earth captured in Figure 3.5 are also meant to be sensational and to grab your attention as it illustrates this particular environmental perspective.

There is a strong ethic within the Doomsday view. The Doomsday ethic is highly focused on the future consequences of our current actions. These consequences appear so dire that the ethic justifies that drastic action must be taken right

FIGURE 3.5   The Doomsday perspective. (Illustrated by Mark Benesh. With permission.)

away. The Doomsday ethic is a very unique perspective in that it is neither nature focused nor human focused. It is crisis focused. Survival is the paramount concern, generally applied to the human race but sometimes inferred to be the earth itself. It is definitely not oriented toward the individual and as such can be very coercive. Individual rights and preferences can justifiably be neglected in the necessary rush to save the world. This view can justify draconian action such as severely limiting births to control population size or restricting gasoline purchases to reign in runaway greenhouse gas emissions from the transportation sector. What a small sacrifice to make to avoid a tragic end to humanity through crushing overpopulation or an overheating planet.

The Doomsday ethic is usually found within larger issues and at large scales. Small local issues typically do not have great enough potential impact to justify the Doomsday judgment. It simply does not make much sense to think that the world as we know it will come to an end because the loggers are clearcutting a mountain forest or because the river that runs through my town is becoming more polluted. The issue must be big enough to justify the approach. Ozone holes, global warming, and species extinction are issues that typify the grand nature of the issue that is required. This tendency for size and scale makes it more difficult to rationally analyze the Doomsday approach to environmental issues. This is because the big issues tend to be the hardest to understand from an individual's perspective. It becomes more difficult to answer the question, "Is this issue really a crisis?"

The Doomsday perspective can be rather insidious as well. It can creep into our minds and alter our thinking. It feeds into knee-jerk emotional reactions and the desire for quick fixes. The Doomsday view can find its way into other complementary environmental views and happily coexist with them. It welcomes the Spaceship Earth or Back to Nature views, for example, and easily shares ideological space with them. It is not too great a mental leap to think that our earthly spaceship is quickly exhausting its critical life support systems, or that our modern technological advances are leading us down a disastrous path. While the attitude that "The End Is Near" is not unique to the environmental movement, it has found a strong foothold within parts of our modern environmentalism.

ANTICONSUMPTION

The Anticonsumption view asserts that environmental degradation is basically a result of our overly materialistic lifestyle. Humanity is simply consuming too much and overwhelming nature's capacity to provide resources and assimilate wastes. We desire too many things, and our desires have too few limits. We have built an economy and a society that is heavily based on consumerism. Figure 3.6 seeks to be the poster child for this view. Let our appetite be basic and frugal and let us pass on dessert.

There is a fairly famous equation within environmental circles that speaks to the Anticonsumption view. This simple equation links environmental impact with human consumption. It is called the I = PAT equation or I PAT (Ehrich and Holgren 1971, Commoner 1972):

FIGURE 3.6 The Anticonsumption perspective. (Illustrated by Mark Benesh. With permission.)

$$I = P \times A \times T$$

where

I = impact
P = population
A = affluence
T = technology

Originating in the early 1970s, the I PAT equation shows that environmental impact (I) is influenced by the human population (P), the level of our affluence or consumption (A), and the technology employed (T) within our societies. It generally states that increasing levels of population, affluence, and technology increase the impacts to our environment. The equation itself is multiplicative rather than additive, implying that large increases in these individual factors could produce even larger impacts to the environment.

Those who are worried about population growth often cite this equation. The concern is twofold. First, increasing human population obviously means that there are more mouths to feed and more bodies to clothe and shelter. More people will require more natural resources for mere sustenance, and more people will generate more waste products that get dumped back into nature. The second concern has to do with our standard of living. More people living at a subsistence level is one thing, but more people living with aspirations of a good comfortable life is quite another. This is the affluence component of the equation and is a strong link between the anticonsumption and population control viewpoints. And again, notice that these two factors are multiplied together to describe their potential impact to the environment. For example, let us assume that our population grows by 25%, our affluence grows by the same 25%, and that technology does not change. The I = PAT equation would predict an increased environmental impact of 56% ($1.25 \times 1.25 \times 1.0 = 1.56$ or an increase of 56%), If we had not given much thought to the I = PAT equation and just how it works, we may be tempted to think that 25% increases in population and affluence would lead to a 50% increased impact (25% + 25% = 50%). The multiplicative nature of the equation tends to predict an impact somewhat greater than we would otherwise expect.

The technology factor in the equation is a bit of a wild card. In one sense, advances in technology permit large changes in how we lead our lives. The development of automobiles and airplanes enabled us to easily travel farther than we ever had before. This development also increased the emission of some pollutants into the atmosphere. Our modern societies have become beholden to oil and petroleum to carry on the normal necessities of life. Some would argue that we even fight wars over oil. We now drill for oil all over the globe from the Texas flatlands to the Arctic tundra to the deep ocean. Our drilling operations are not perfect, and every so often we seem to have a major spill that pollutes the local environment. So, in this example, advances in technology have resulted in increased environmental impact.

The other side to the technology argument is the case where advances in technology result in less environmental impact. Using the same transportation example, technological advances have improved the fuel efficiency of cars and planes and reduced the amount of pollutants per car or per plane emitted into the air. The automobile of the 1950s is a far cry from its descendant of the new millennium. By most measures, styling and nostalgia aside, our new modern cars are far better than yesteryear's models. Yet, at the present time, most autos still consume fuel and create emissions, emissions that would not occur without the technology that allowed the creation of personal automobiles in the first place. Perhaps someday we will see electric or hydrogen-powered cars that are fueled wholly by wind or solar power. Forgetting for a moment any environmental impact associated with their manufacture, perhaps this technological leap will allow us to forgive technology its impact upon the natural world in this case.

The Anticonsumption view maintains that if we can control our level of consumption, we can positively impact our environment. Of the I = PAT factors, affluence is the element we can most directly and humanely control. The current human population of the earth is already with us and is not easily changed in the short term without harsh and extreme action. Technology advances at its own pace and in its own direction. While we can influence it through our choices in funding of research and development activities, we cannot control it nor can we predict with much certainty if and when a new technological breakthrough will occur. The Anticonsumption view asserts that consumption and affluence are the keys. It hopes that our population will not expand greatly, and it hopes that beneficial technologies will come forward that reduce our impact on the earth, but it relies on our collective willingness to live within the earth's means as the primary path to lessen our negative environmental impacts.

The Anticonsumption view desires that we limit the environmental impacts of our necessary consumption while at the same time limiting our unnecessary consumption. We should strive to minimize the waste and pollution that accompanies our necessary activities while we also minimize our excess. Does every family need a large and spacious home, as well as three or four automobiles, complete with a heated and air-conditioned garage to hold them? Do we need a television in every room of our homes, and do they all need to be on at the same time? Do we really have the liberty to throw aside items we no longer desire and bury them in the earth? The Anticonsumption view prescribes answers in the negative to these questions. This very frugal perspective might be seen by many as draining much of the fun and enjoyment out of life. Knowing our own human nature, we should recognize this perspective as being very difficult to market. The Anticonsumption view maintains that it is ultimately necessary, however, and ultimately beneficial to humanity in general. The luxuries we discard may allow another to live a decent life, humble by our standards maybe but complete with the basic necessities to survive and be happy.

The last environmental perspective we will discuss in this chapter is a very happy one indeed. It does not get a lot of attention in our current environmental discourse but is certainly present and worth our examination.

CORNUCOPIA

The Cornucopian view is a very positive and optimistic view of our future with respect to nature and our environment. There are no environmental catastrophes looming in the near or distant future. Humanity possesses the intellect and ingenuity

FIGURE 3.7   Horn of plenty (cornucopia). (Illustrated by Mark Benesh. With permission.)

to provide solutions to the environmental difficulties we may encounter. The future is truly bright, and as Figure 3.7 implies, there is plenty to go around.

This view looks at our human history and sees an evolution of progress socially, economically, and environmentally. It points out that our standard of living has steadily increased over the decades and centuries. It recognizes that human life span and general health has increased as well. It looks at technology and sees how it has drastically improved our lives and given us new and more efficient means for communicating with one another, for transporting ourselves to places near and far, and for producing the goods and services we all need and enjoy.

The Cornucopian perspective includes a basic trust in the combination of humanity and technology to effect necessary environmental change. It is a human-centered ethic with an appreciation for nature, but one that does not worry too much about negative environmental events. Humanity will find a way to fix the problem, our economies will continue to grow, and our societies will continue to prosper. The Cornucopian view does not suffer coercion. It does not need to, for it believes in the ever-progressive unrelenting nature of human beings to continuously improve their surroundings. This view holds at all scales and easily accommodates the regulatory process as an effective way for us to manage our environmental activities.

This view does acknowledge environmental issues as being important, and as such is significantly different than the Nonbeliever view discussed earlier. The

Cornucopian view, however, simply does not see catastrophe or hardship associated with the issues. It sees a new challenge and envisions humanity rising to the challenge. It has great faith in the ingenuity and creativity of the individual and society. It looks at global climate change and believes that we will find a way to fix the problem or adapt to its effects. It sees a future where all will have access to clean drinkable water and healthy food. It believes that we will find a suitable energy source to power our economies far into the future. It asserts that things are already getting better.

The Cornucopian view proudly points to the past and says, "See how far we have come already." There is little doubt within this view that humanity will continue to improve its surroundings and its quality of life.

Hopefully, this short list and descriptions of a few perspective-based ethics, as well as the previous discussion of issue-based and process-based environmentalism, is instructive in organizing our thinking around environmental issues. These categorizations may overlap at times, too, as we move from our own individual perspective about the environment through the environmental issues themselves, and finally to a preference for how we respond to the issues. Hopefully, this discussion will not only inform our own environmentalism but also make it easier to communicate with others who see the environment differently from us.

Let us conclude this chapter with the promised detailed discussion about the ethics of practical environmentalism. It is basically a process-based ethic with just a smattering of perspective-based ethics mixed in. As you will see in the chapters to follow, there is a set formula, an established pattern of analysis to follow as an aid in making environmental determinations and gauging environmental responses. The process itself is defined around the four pillars incorporating both nature and humanity into the method. The process is straightforward and relatively easy to understand. It can be applied broadly across the range of environmental issues and so qualifies as a process-based ethic in this regard. In fact, broad application of the method to the great range of environmental issues we face is definitely a goal of practical environmentalism.

The smattering of perspective within the ethic comes from the idea of practicality. Practical environmentalism seeks actions and solutions that individuals are able to employ and put into practice. It accepts small efforts by you and me, and celebrates our small successes. It holds the perspective that small environmental changes in the "right" direction are valuable and recognizes that many small actions taken together can produce significant change. Practical environmentalism does not seek perfection and accepts that we will always exert some negative forces upon our world. The biblical Garden of Eden does not exist, but we can and should live within earth's capability to support humanity, now and in the future.

As mentioned previously in this chapter, the ethics of practical environmentalism are based on the independence of the thoughtful individual. It relies strongly on the willingness of people to examine the issues and their choices with respect to the issues. It suggests a rational framework, out of practicality, to be employed to guide us to ask appropriate questions and to make appropriate comparisons. It takes no particular stand on environmental issues per se. It is not issue driven. Instead it requires us to take a holistic approach to incorporating environmentalism within our daily lives.

As touched upon briefly in Chapter 1, practical environmentalism takes a generally deconstructionist view of an environmental argument or issue. I use the term "deconstructionist" rather loosely and apply it to describe the method of breaking an argument apart and into smaller more manageable pieces for analysis. Deconstructionism is more properly known in the study of philosophy as a way to examine the meaning of texts as proposed by Jacques Derrida in the late 1960s (Derrida 1974). In very basic terms, Derrida argued that there is no single or simple way to understand the meaning of a written work. Indeed, there may be many valid and worthwhile, even contrary, interpretations of the same text. Practical environmentalism accepts this basic premise. It accepts that environmentalism is neither simple nor singular. Deconstructing an environmental issue into the four pillars is an implicit acknowledgment of the complexity of environmental issues. Focusing on individuals' choices and opinions values the multiplicity of perspectives and allows contrary viewpoints to fit nicely within the approach. Deconstructionism is a main component of the ethic of practical environmentalism.

Practical environmentalism can be employed at essentially any scale, although it is probably best at local scale. Here, in our daily lives and in our hometowns, the process of practical environmentalism is designed to structure our thoughts and our analyses to produce meaningful environmental benefit to ourselves. Thus, particularly at the local scale, practical environmentalism is also very human centered, with the human being you! The ethic includes the importance of the individual, both in ascertaining appropriate action and in defining benefit and success.

Being human centered and individualistic, practical environmentalism tends to minimize the impact of coercion that is sometimes present in other environmental approaches. It focuses on actions that originate from within the individual instead of actions that may be forced upon us by other groups. This is a key differentiation concerning practical environmentalism. Many other forms of environmentalism, whether they be regulatory, protest oriented, or issue based, often rely on some type of coercion to drive environmental action. As noted before, regulatory action is by definition coercive through laws enacted and enforced by governmental bodies. Protest- and issue-based environmentalism can be coercive through the methods used to gain attention to their claims and the manner in which ideas and solutions are marketed to us. With practical environmentalism being founded on the individual, even driven by the individual, it is very difficult for coercion to take hold. How can we force ourselves to do something we do not wish to do?

In general, and seemingly out of some necessity, our modern environmentalism is rather coercive. It seems far more interested in convincing us instead of informing us. Rarely are we presented with a range of options to cure some environmental ill. Rarely are we asked to consider any impacts beyond the environmental issue itself. Perhaps individually we have allowed it to be this way because it is simpler and easier to "go along" rather than invest the time and effort to become better informed and better able to identify and determine appropriate responses to environmental issues.

Coercion within the environmental movement is made easier by the manner in which environmental issues are presented and debated within our society. Confusion and misunderstanding are often heightened by incomplete and inadequate communication of the issues, impacts, and potential responses. Chapter 1 mentioned several confounding factors that tend to compound our difficulty in analyzing the extent of the environmental threat and the "best" response to the threat. The next chapter discusses these confounding factors in detail. These factors are the product of our history and our ethics with regard to the environment, and a better understanding of their existence serves as a solid foundation for the implementation of a more practical environmentalism.

REFERENCES

Lovelock, J. 1975. The quest for Gaia. *New Scientist.* February 15, 1975.
Ehrlich, P.R. and Holdren, J.P. 1971. Impact of population growth. *Science* 171: 1212–17.
Commoner, B. 1972. The environmental cost of economic growth. *Population, Resources and the Environment.* Washington, DC: Government Printing Office. pp. 339–363.
Derrida, J. 1974. *Of Grammatology.* Baltimore, MD: Johns Hopkins University Press.

4 The Confounding Factors

As mentioned in the introductory chapter, the confounding factors are elements within our environmental decision making that make the process more difficult and less likely to yield beneficial results. Once again, these factors are

- Competing objectives
- Uncertainty
- Measures of success
- The fallacy of prediction
- Assumption of future states
- The problem with percentages
- A history of paranoia
- A crisis mentality

These eight are certainly not an all-inclusive list, but they are representative of many of the troubles we encounter when we try to solve environmental problems. Some are problematic because of differences in values and opinions between people. Some are troublesome because of the tools we use to approach environmental issues. Some are ingrained within our history and the way our society has reacted to the growing environmental movement. Let us attempt to shed some light on these eight and see how they influence our environmental decision making.

COMPETING OBJECTIVES

Competition seems to be the norm in our modern societies. It has probably always been the norm as humans have struggled for survival. In our prehistory, we would compete with the elements, with our prey, and even with each other for the food and shelter we required. Now we compete for the jobs that provide the currency we require to buy our groceries and pay our mortgages. For many of us, before we are able to compete for our employment, we compete for our schooling. We strive for acceptance into colleges and universities, and once there, we compete with ourselves and with our classmates for the marks in our exams, the grades in our classes, and finally our diplomas and our standing among our peers. We hope that our résumés will impress the recruiter during the job interviewer and that we will get the offer. Or perhaps we are the entrepreneurial type and we compete for the bank loan to get our business off the ground or the infusion of capital from investors necessary to expand and grow our business. In the world of business, the products we manufacture and the services we offer compete in the marketplace. If these products and services fail in the competition for customers, our livelihood is placed in jeopardy. If they succeed, we hope to become wealthy and comfortable, maybe even happy.

Competition is part of the American ethic. Our society expects competitiveness. Not only is it ingrained within our business models, but it infiltrates our leisure time as well. We now have "reality" TV shows that fabricate a competition just to amuse us and generate ratings and advertising revenue, of course. Our news broadcasts tell us which new movie outsold all others at the box office this week, or how many albums have been sold by our favorite pop star's latest effort. We can watch the Oscars, the Emmys, or the Grammys among a host of other awards shows to see which movie, television show, or song is the "best." We have designed such competitions to be television galas in and of themselves. I wonder when we will see an awards show for the best awards show.

Of course, competition is obviously present in athletics. This should not be a great surprise as sports are competitive by definition. Professional competitions exist for baseball, basketball, football, golf, hockey, soccer, and tennis. Weekend television is filled with these events and also with pre- and postgame shows dedicated to the analysis of the event. I am a baseball fan myself. Growing up in a near-north suburb of Chicago I am a loyal fan of the Chicago Cubs. I have many fond youthful memories of summer afternoons spent watching the Cubs play ball on TV or occasionally watching in person at Wrigley Field. The Chicago-based television network WGN showed just about every game back then, so it was easy to follow the team even if you could not get to Wrigley. I remember being very happy when the Cubs won and very disappointed when they lost. I am a Cubs fan today even though I do not reside in the Chicago area anymore. I have grown to understand baseball on a deeper level, and I can appreciate fine play even by the team that is playing against the Cubs. Still, I would really prefer for the Cubs to win.

I recognize now that as a child I was competitive. My children today are competitive, too, by varying degrees as part of their nature. To a certain degree, competition can be a healthy motivating factor when our children are mature enough to understand it. Perhaps though, we push a little too hard at times. Competitive athletics for little 6- and 7-year-old children might be a bit much. The harsh realities of getting cut from the high school sports team or not getting a good part in the school musical come soon enough.

Competition is part of our society, part of us, and present in the environmental field as well. Governmental environmental agencies compete for their operating budget among all the other agencies within the government. Environmental programs compete with one another within environmental agencies. State agencies compete with national agencies over jurisdiction and emission limits. Private environmental action groups compete with public agencies and private corporations, sometimes to the point of legal action.

Competition is even present when we have already committed ourselves to some course of environmental action. For example, let us suppose that we desire to reduce air pollution from industrial facilities. We have decided that the noxious chemicals spewing forth from tall smokestacks has gone on far too long. Let us also make the very fair assumption that shutting down these industrial facilities is only acceptable in extreme cases and that most facilities will still find profits in continued operation even with additional pollution control equipment installed in their plants and added to their list of depreciating assets. Therefore, we attempt to institute rules and regulations requiring pollution control equipment to be installed in these industrial facilities. In our attempts to remove the offensive pollution from the air, we must realize that it still has to go somewhere. If we are successful in removing it from the stream of gases going up and out the smokestack, where do we put it? Perhaps we filter these stack gases through an aqueous solution in a wet scrubber and we transfer the pollutant to the facility's wastewater stream. Perhaps we filter the stack gases through a baghouse or an electrostatic precipitator and transfer the pollutant in solid form into ash that finds its way to a local landfill. We have cleaned up the pollution in the air but imposed a burden of pollution to our water or land. On balance this may be the proper way to treat the pollutant, but it must be remembered that removing a hazard from one place often inserts it into another. One person's objective to improve air quality may compete with another's objective for pristine water or for the elimination of landfills.

The foregoing example is admittedly simplistic. It functions under the assumption that the noxious chemicals flowing up the smokestack are bound to exist. It neglects the opportunity to change the form of the pollutant itself through some treatment process—biological, chemical, or combustion reactions, for example—that would yield a more benign compound. We could also consider changing the manufacturing process itself to one that does not produce these nasty compounds in the first place. Changing a manufacturing process is usually not an easy task though. Even if the proposed new process makes it from the research and development labs through product manufacturing trials and proves to be possible, it is probably going to be expensive.

In these scenarios we have reduced the pollutants in question but may have created others in their place. In the cases considered, whether we filter out the pollutant or modify its form, we have probably also added complexity to the operation and increased the amount of energy (fuel and/or electricity) required to operate the process. These additions have environmental impacts of their own as well. Remediating old pollution competes with creating new. Investing in new manufacturing processes competes with investing in pollution control equipment for old manufacturing processes.

Competing objectives are ever present in our environmental decision making at some level. There are usually trade-offs to be made, even if you are trading one positive result for another. Rarely are we all in perfect agreement about our environmental objective or the method of its achievement. A decent understanding and appreciation of the competing objectives at play within our environmental debates is very helpful in facilitating the success of our endeavor and the betterment of our environment.

UNCERTAINTY

Uncertainty is particularly problematic in the environmental domain. Our natural ecosystems are wonderfully complex from a purely scientific point of view but notoriously difficult from an applications perspective. Our environment is a system with various inputs, outputs, and forces acting upon it. It is formed where biology, chemistry, commerce, culture, ecology, human nature, industry, physics, and zoology all intersect. It is a very complex system, and it can be very difficult to predict how the system will react when a change is made.

We human beings hate uncertainty. Knowledge, understanding, and security make us feel very comfortable. It is so much easier to make decisions when we have all the facts. It is hard to factor uncertainty into our thought process. It requires a level of thinking that can get very complicated and a lot of mental energy. We must acknowledge, however, that uncertainty exists in our environmental decision making whether we like it or not. Ignoring uncertainty is a fantasy at best and a significant mistake in many instances.

Perhaps when we are faced with an environmental dilemma and we do not have a perfect understanding of all the forces at play, we should prefer to make small changes to ensure that the results are at least in a generally positive direction. This is basically the concept of "Adaptive Management" in that we make a change fully expecting that the result will not be exactly as predicted. The modern development of this concept in the ecology field is generally attributed to work in the 1970s by C.S. Holling and Carl J. Walters at the University of British Columbia. Adaptive management requires that we monitor and analyze the results from our actions and incorporate the new information into our decision making before taking additional steps or making more changes. We expect to learn something and we expect to improve our ability to predict and manage the system under study. Adaptive management is especially useful in situations where we cannot conduct a formal experiment, where we cannot reduce the problem to a set of test tubes in a laboratory.

In reality we practice adaptive management all the time. Good students improve their study skills over time as they learn what approaches work best for them. Our approach to parenting often changes with a second or third child that enters our family. Not only have we learned something from the first child but we also learn that all kids are different and that what motivates daughter number one might not do anything for daughter number two. It is no great leap to think that adaptive management might also work very well for environmental issues.

Another aspect of uncertainty has to do with an idea called the "precautionary principle." This is the "better safe than sorry" approach. The reasoning is that even given uncertainty, given the fact that we do not know exactly how serious some issue is, or how devastating some change might be, we should take some action to avoid the future crisis. For example, let us assume that a small insect, perhaps a species of beetle, is found to be extremely susceptible to a particular fertilizer that is widely used in farming. In fact we have recently learned that this fictitious bug is dying in record numbers in its habitat that borders farm fields. While this beetle is not considered to be a major pest, it is not considered to be particularly valuable to us either. We know of no important benefits that this species brings to humanity other than it exists within the ecosystem and is eaten by some small animals. The precautionary principle would urge us to take some action to protect this organism even though we do not really know if there would be any ill effect if it were to vanish from nature. We might restrict or regulate fertilizer usage in some areas in an attempt to safeguard this creature, for we really do not know what might happen if we were to drive it to extinction. Might there be some ecological crisis if it were removed from the food chain? Is it really performing some useful function that we are simply not aware of?

The precautionary principle is becoming very well accepted worldwide in government and environmental advocacy organizations. The inherent problem with this approach is that it is often applied with regard only to the perceived environmental crisis and without sufficient thought to the impacts of the proposed remedies (hence the need for the pillars). While the precautionary principle incorporates uncertainty and is very useful as a motivating factor, we must be careful not to overextend its value in using it as a decision-making tool.

MEASURES OF SUCCESS

Success is an elusive target. It is very personal, and there is no standard for uniform application. For too many around the world, success is having enough to eat or a safe comfortable place to call home. Some think success is measured in monetary terms. For others it is power and prestige, or perhaps fame. Our definition of success tends to change throughout our lives. In our high school years it may be as simple as getting a date with the attractive boy or girl that sits next to you in class, or making the varsity basketball team or being in the top 10% of your class academically. Later on, as the letter jackets become dusty and our high school memories begin to fade, success might be a good job, a successful business, a loving spouse, or a vacation home in the mountains.

Since success is so personal, it is a bit tricky to apply it beyond ourselves. It tends to become our perception of events happening around us where we have little direct control. A successful election is one where your preferred candidate wins, or maybe where the candidate you despise loses. A successful baseball game is one where the Cubs win—that is *my* definition, of course. A successful baseball season is another matter. Certainly, a winning record is required to be successful, but what about making the playoffs, or winning a league championship or even the World Series? Success here definitely depends upon the individual. For a Yankees fan, I expect success often comes only through World Series championships.

Success in environmental matters is also something that is typically far beyond our direct control. It is our perception of the state and well-being of nature. Not only do we all have different perceptions of nature but different understanding of how nature functions. I daresay that none of us has a perfect understanding of how nature works. So not only do we perceive nature differently but the degrees of our misunderstanding of nature varies as well. The picture of our earth that we present is our own but undoubtedly somewhat flawed.

Because we all have different definitions of success, different perceptions of the natural world, and different understandings of just how nature behaves, measuring success in environmental matters can be particularly problematic. Some of us might define success as avoided environmental catastrophe, while others might require an earthly utopia where environmental degradation in any form ceases to exist. What is good enough for me might not be good enough for you. It is of course a matter of degree, but the range of possible success measures is very large. This presents a situation similar to what we find with the problem of competing objectives in that facilitating environmental progress requires an appreciation of each other's definition of success.

The range of success measures is especially evident when we consider national regulations on environmental quality. For example, how much mercury vapor in our air is acceptable? Is 10 parts per billion (0.0000001%) OK, or is zero better? How low a concentration can we really achieve? How low a concentration can we really measure? How much money are you willing to spend to reduce the concentration from 10 parts per billion to 5 parts per billion?

Success within environmental matters can be elusive. Political definitions of success may differ markedly from scientific analyses or personal values. It might not be a question of right or wrong but instead a question of what is acceptable to the majority. It might become an issue of what is realistically achievable. It can become more a matter of judgment and less a matter of fact.

FALLACY OF PREDICTION

The fallacy of prediction refers to our gross inability to predict long-range future occurrences with any degree of success. This happens over and over again, and has been going on this way for centuries. Garret Hardin, in his essay "Tragedy of the Commons," predicted a future of mass starvation and poverty because of humanity's inability to produce enough food to keep up with our exponentially growing population (Harden and Baden 1977). In the early 1970s, the Club of Rome, a group of influential scholars and thinkers, predicted a similar fate (Meadows et al. 1972). Also in the 1970s there were warnings of a coming ice age, and the 1990s saw the ozone hole take center stage. Currently we are inundated with dire predictions of global warming.

Long-range, large-scale predictions are usually wrong, and the percentage of really wrong predictions goes up as we extend the time frame. The real question is how wrong any individual prediction will be. The fallacy of prediction occurs because we typically discount the likely error in the prediction and instead focus on the difference between the prediction and our current state. The error exists because our knowledge and understanding of complex environmental issues is imperfect, some might even say flawed. But the error in prediction is not necessarily the same thing as a mistake. Certainly a clear mistake in theory or programming could deliver an unreasonable and unrealistic prediction, but so also could a simple lack of understanding, an incomplete level of information, or some unaccounted factor that has or will have a major influence.

Yet, prediction is basically the only tool or method we have to consider the future, and in some cases we can apply it relatively well. We can accurately predict the time the sun will rise on any particular day. Unfortunately, we cannot predict very well if we will actually be able to see it or if it will be hidden by clouds. We can expect that, on average, a baby will be born nine months after conception, but we cannot guarantee that any particular baby will be born then. Some come early, some very early, and some a little late.

We tend to forget these lessons about how well we can really predict the future when it comes to the environment. Now, with the benefit of very fast computers and adept programmers, we use complex computer models that produce volumes of official-looking reports that portray our future with convincing certainty. All the scientific formulas and advanced mathematical equations make it appear more like a calculation than a prediction. Yet, it is still a modern-day divining rod, even with all the knowledge and education that we pour into it. It can still be wrong, easily wrong, even though it looks right, so neat and clean. So, acknowledging that prediction is a necessary evil in dealing with our future circumstance, we should consider treating it much like the precautionary principle and realize that the art of prediction is best left as a motivational factor and not as a decision-making tool.

ASSUMPTION OF FUTURE STATES

The assumption of future states is closely tied to the fallacy of prediction because it is a direct result of the prediction itself. The problem here is that there is not just one future state but a whole range of possible states exist, each with different probabilities of occurrence. In effect it is a probability distribution. It is statistics. Not many people relish the art and science of statistics, so it is easy to understand why a single future state is assumed. It is much easier on the brain that way. It is also essentially wrong, just as predictions are essentially wrong.

The assumption of a single future state often occurs because it is easy and because comparison with our current condition is then simply a comparison of two states. This is perfect for news headlines and 30-second commercials on TV. Some have predicted a future state where San Francisco and New York are under water due to the rising sea levels forecast by the global warming theory. That image is certainly too compelling for the media to resist. Never mind that there are a myriad of other future states that also fit within the global warming theory. The single-future-state assumption allows us to keep our thoughts at an absurdly simple level. We do not have to think too much that way. All that is then required of us is to hear, accept, and obey.

Another problem with our assumption of a single future state is that it discounts actions that may mitigate a problem or change direction entirely. It discounts the forces, both natural and social, that tend to keep things in balance. For example, say, a biologist has noticed a dramatic increase in the deer population in a particular region. Let us assume that it has doubled over the course of 5 years. This could be presented as a major problem: "In ten years the deer population will be four times what it is today and eight times what it was 5 years ago." The headline might read, "DEER POPULATION TO GROW 400%!" This prediction identifies only one future state and neglects all the forces and actions that will likely occur through time to challenge this particular outcome. Hunting may increase. Disease, starvation, and death caused by collisions with vehicles may increase. Deer may migrate in search of food. All these possible actions are likely to dampen the forces causing deer population to grow. The assumption of a singular future state is a disservice to the problem or question at hand. It is really a misrepresentation and can often lead to undesirable actions.

THE PROBLEM WITH PERCENTAGES

The problem with percentages is that they are so darn hard to understand, and they tend to draw attention away from the basic issue at hand. Take for example the statement "Government reports indicate the economy grew at a robust 2% in the preceding quarter." That sounds good, right? But what does it really mean? In this particular instance it is hard to know for sure. First you would have to understand how economic growth is measured, and then you would need to know the basis of comparison for the 2% value. Is the 2% compared to a previous quarter or possibly the previous year? Percentages are very good at summarizing information without necessarily giving you the knowledge you need to truly understand the information. This works well if you are trying to convey a message, but maybe not so well if you are trying to fully inform people so they can make educated decisions.

It is not that the concept itself is so difficult. A percentage is essentially a ratio after all; you simply divide one number by another. The problem is that there are lots of different percentages that we can talk about. Percentages are used to talk about a portion of a whole. For example, how big is your piece of the pie? Percentages are also used to talk about how much something is changing. We regularly use percentages to describe the change in the cost of living index, for example. It can get really silly when some news anchor starts telling us about the decrease in the increase of some critically important statistic, say, the country's gross national product. Suddenly the percentage itself has become the news, and we tend to forget about what we are trying to measure in the first place. All we hear is the downward or negative percentage, when in reality the thing we are measuring, the national economy in this example, is still growing, just more slowly than before. Is that necessarily a bad thing? When did it become necessary that each year must outperform the year before? Why do we expect our stock portfolios to grow by greater and greater percentages every period? Why must corporate profits increase by continually increasing percentages? We have become dulled to reality as we are force fed a steady diet of obtuse percentages that we willingly ingest in the few moments we take to consider the issue at hand.

Another problem with percentages, particularly in the environmental field, is when we start setting goals and making plans based on percentages. This is very common in discussions about air pollution and climate change when we hear things like "Our goal is to reduce emissions by 20% by the year 2020 compared to 1990 levels." That might be just the first step. We might have a 50% reduction by 2050 as step number two. Why 20%? Why 50%? Well it is kind of catchy to say 20% in 2020 and 50% in 2050. That is a smooth-sounding slogan that will play well on the evening news. I wonder if the silver-tongued devil who thought up the slogan has any idea if it is realistic, or even possible! Large percentages sound so grand, so noble. A 5% reduction is so puny, so unremarkable. Would a 5% off sale entice you to go shopping?

HISTORY OF PARANOIA

Paranoia is another phenomenon that is often present in environmental issues. Sometimes it seems to be so deeply ingrained within us that it naturally leaps to the forefront of the discussion. Perhaps it is a natural result of humanity's evolution through the centuries. There is always something to worry about and something to fear. It is easy to comprehend how this outlook could be very beneficial in past societies where attack, disease, starvation, and death were too often present. There was something to fear then, and paranoia was probably an important survival mentality. Things are different now though, at least in many places around the globe.

Admittedly, there are places, in nations rich and poor, where some people do indeed have reason to fear for their survival. I would argue though that these situations are rooted in social circumstance and need not represent the typical condition of our society. These situations are a failure of our societal ethic and not the norm. We possess the knowledge and technology, if not the will, to provide sustenance and safety to the people of the earth.

Paranoia finds a comfortable home within the environmental movement. We read the news and hear the expert opinions with the nagging feeling that something really bad is just around the corner. Perhaps some evil force exists within greedy corporations, intent on exploiting the environment for shareholders' gain, not to mention year-end bonuses. Maybe corrupt governments are willing to compromise the environment to advance their political agendas. Perhaps God is fully frustrated with our failings and is willing to unleash terrible calamities not seen since Old Testament days. It is so easy to expect the worst, and this makes us very susceptible to the prediction of crises to come.

CRISIS MENTALITY

A crisis mentality refers to a state of mind or a type of communication where the issue at hand is portrayed in the direst terms. It can be considered to be the fraternal twin of paranoia. These two commonly negative phenomena are mutually linked and supportive. They often exist in tandem and reinforce each other. A new environmental issue can be portrayed as a crisis because our history and acceptance of paranoia allows it. A nearly hysterical media provide the latest environmental crisis to millions through 30-second TV news spots or terse, teasing Internet headlines that attract our attention but fail to inform. The crisis mentality packs a powerful emotional punch and grabs headlines and ratings. It makes for "great" TV.

The problems with this approach are obvious upon the least reflection. Environmental issues are too complex, too subtle, and too long-lived to be resolved under the intense pressure and shortsighted reasoning that comes with a crisis. Pearl Harbor was a crisis. The stock market crash of 1929 was a crisis. The loss of the Titanic was a crisis. The bombing and isolation of Great Britain in World War II was a crisis. Crises demand swift and immediate response. The danger is immediate, and quick action, even if not perfectly conceived, is warranted based on imminent potential harm and loss of life. We willingly sacrifice the most thoughtful response in order to deliver an appropriate response as fast as we can.

Most environmental issues are not crises simply because they take a long time to develop. Their impacts tend to be slow and subtle, and our scientific knowledge is often too limited to truly know what is going on. There are a few notable exceptions, of course, some of which will be mentioned in the next chapter, but in general I believe the assertion that environmental issues do not qualify as crises to be fair. The danger in treating each environmental issue as a crisis is analogous to the old fable of the boy who cried wolf.

Are we already becoming desensitized to the environmental alarms? Are we growing tired of all the threats and dire predictions that seem so far off and distant from our current lives? Will we be asleep at the switch if a truly harmful environmental calamity arises and perhaps suffer the one-in-a-million train wreck that truly is a crisis?

Environmental decision making is hard enough without artificially constraining our thought processes with the shackles of these eight confounding factors. And realistically, there is probably several more I failed to include. The point of this chapter is that acknowledging their existence will ultimately improve our ability to act in ways that will improve the environment. As we learn to better recognize these factors, we can begin to strip away the layers of confusion and start to base our decisions on the factors that really matter. The rest of this book will focus on the factors that really matter—the four pillars.

REFERENCES

Hardin, G. and Baden, J. 1977. *Managing the Commons*. San Francisco: W.H. Freeman.
Meadows, D.H., Meadows, D.L., Randers, J., and Behrens, W.W., III. 1972. *The Limits to Growth*. New York: Universe Books.

5 Environmental Degradation— The First Pillar

Environmentalism today is mainly a reflection of perceived environmental degradation. The modern form of environmentalism that exists within our minds and spurs our actions is typically a reaction to some sort of unjust despoilment of the natural world. Our modern environmental movement does not require that we experience this degradation firsthand. It does not require that we walk the beach to see the crude oil wash ashore from the supertanker that sprung a leak. It does not require that we witness the rifle shot that killed one of the few remaining dodo birds. It does not require that we climb Mount Kilimanjaro each year to record the retreat of the glacial ice. All that our modern environmentalism requires is for us to develop a mental image of the earth suffering and that delivers the basis for environmental judgment and response.

Environmental degradation within the framework of Practical Environmentalism is substantially different. Environmental degradation is critically important and the first pillar of practical environmentalism, but it is not the only factor. The other three pillars are important, too. While this chapter is devoted to discussing environmental degradation and how it fits into the four pillars of practical environmentalism, the chapters that follow will discuss the other three pillars in similar detail. In each case we will describe and discuss the pillar itself and investigate a few important criteria and issues relating to that pillar. Following that we will bring all four pillars back together again so that we can render an enlightened environmental judgment.

Within practical environmentalism, "enlightened environmental judgment" is achieved by a relatively simple scoring system that allows us to transform our reasoned opinion of each pillar into numerical values that can be combined to inform our actions. At the end of this chapter I will present the scoring system used to judge the severity of environmental degradation. In succeeding chapters we will apply the scoring system to the other three pillars. This will be the "pillar analysis" that directs our action toward environmental betterment. Before we get too far ahead though, let us focus on the first pillar.

Historically, environmental degradation has been the main focus of environmentalism. At its core, environmentalism has been a reaction to environmental degradation. From what I now see on television or the Internet, from what I read in newspapers and magazines, or from what I hear on the radio, our society's perception of environmental degradation is that it is becoming more and more serious and it is developing far-reaching and potentially catastrophic implications. The general sense is that our environment is getting worse. It is kind of hard for me to fathom that things are getting worse when we as a society have invested so much time, effort, energy, and money into environmental legislation, regulation, remediation, and education. Have all our efforts over the past few decades been in vain? Have we merely slowed the pace of the inevitable destruction yet to come? Personally, it is difficult for me to accept these assertions, but that is often what we hear. Let us begin to put environmental degradation into perspective, both as the underlying theme of our modern environmentalism and as a fundamental component of practical environmentalism. Permit me to begin with a short story from my own experience.

When I was in graduate school, I had the opportunity to be a teaching assistant in a large environmental studies course. I taught two or three classes a week, composed mainly of freshmen and sophomores. I considered it part of my job to stir them up every once in a while and so, at some point during the semester, I would inevitably state, "Ninety percent of the claims of environmental degradation are simply not true. The real trick is figuring out which 10% are."

I made this claim merely to make a point and hopefully to start them thinking that particular day. That, mind you, was no small challenge on some early Friday morning classes. The point, of course, was to get them to think independently instead of merely ingesting and regurgitating every environmental calamity that was thrust before them. The 90% figure was merely a guess on my part as I had done no research as to the historic percentage of environmental claims that ultimately proved to be true or false. I picked 90% because it was a high number, and I expected it to directly contradict my students' expectations. I hoped they would disagree with me and begin to think of ways to voice and explain their belief to the contrary. I also wanted my students to understand that many environmental claims are conditional and based on predictions of future events that may or may not occur. True environmental harm does not occur in the future; it occurs in the present, a present that may be heavily influenced by activities from the past.

There is little doubt in my mind that dangerous environmental degradation has existed in the past and could continue to exist in the future. There is significant potential for bodily harm, sickness, and death when potent biological, chemical, and physical agents are improperly released into the environment or improperly safeguarded from human contact. History has shown us that dangerous conditions can be brought upon an unsuspecting populace. Cholera outbreaks in nineteenth-century Europe; chemical poisonings in Bhopal, India, Love Canal, New York, and Times Beach, Missouri; and radiation exposure at Chernobyl in the Ukraine are all vivid examples of environmental degradation affecting people. People have died, both quickly and over the course of years, from the extreme degradation of their local environment.

Yet all environmental issues do not carry with them the same potential for harm, nor do they affect the same groups. There is a priority of issues, a logical hierarchy of environmental concerns that we face. The most acute environmental concerns are typically local and have a direct bearing on the health of people. It is really hard to ignore a major chemical spill or radiation exposure that kills or cripples many people very quickly. In such situations you would be lucky to get a few minutes' warning from an evacuation siren or perhaps from a police cruiser as it hurries through your neighborhood downwind of an accident site. You might have enough time to gather your family, perhaps your pets, too, and escape before you succumb to this inhospitable environment. The environmental degradation is obvious even if short lived. It is tempting to think of these occurrences as accidents instead of environmental degradation, but the consequences can be so dramatic and so severe that I believe they are worth including. I would argue that acute environmental degradation via "accidental" happenings deserves at least as much attention, and in reality probably more, precisely because of the acute and dramatic consequences.

First and foremost, the environment is the air we breathe; the food and drink we ingest; and the physical, chemical, and biological forces that impact our bodies. It is immediate and ever present. Every few seconds my body desires another breath. I cannot wait very long for the air to clear or the toxic cloud to dissipate before I must draw the air from in front of my face deep into my lungs. I really have no choice but to accept whatever quality of air that exists from moment to moment and is available to me. I can survive for a few days without drinking and many days without eating but only a few precious moments without breathing.

I must also protect my body from the hazards of nature in order to survive. Extremely hot or extremely cold conditions can rob me of life long before my body succumbs to thirst or hunger. Proper clothing and secure shelter is an absolute necessity in the parts of our globe that experience extreme temperatures, violent storms, or occasional flooding. We are constantly interacting with our local environment, and generally do pretty well. Our environment is typically benign and generous in its offering of air, water, food, and shelter. "Accidents," though, can change this very quickly and degrade our local environment drastically enough to put us in harm's way.

There is a very big environmental accident occurring even as I write these words. There is a vast amount of crude oil seeping into the Gulf of Mexico south of the Louisiana coast from an accident at an offshore drilling rig that also tragically claimed the lives of 11 workers. This oil has been spewing forth from the ocean floor since an explosion rocked the drilling rig. As you might expect, the accident and spill has garnered a large amount of media coverage and public outrage. Besides killing 11 people, the explosion apparently also caused the drillers to lose control of containment of the well they were developing thousands of feet below the surface. As a result, millions of gallons of black murky oil escaped its subterranean confinement and spewed upward into the Gulf waters. As it rose it mixed with the currents and by the force of its own dispersion to wash up on our sun drenched beaches. We see pictures of seabirds coated with oil and struggling to raise their wings. We wonder about this oil coating the fertile undersea beds of shrimp and oyster and creating economic hardship for those who fish these waters. This oil and the energy it contains was supposed to eventually find its way into the gas tanks of our cars, trucks, and SUVs so we could commute to work or run out and go shopping. A large oil company is being blamed for the environmental catastrophe, and I am reminded of another great spill in the Gulf of Alaska when the supertanker the *Exxon Valdez* ran aground, belonging to a different large oil company by the way.

This is a very public and emotional display of environmental degradation. Hopefully, it will be short lived, but it is probably too early to say that right at this moment. What is very interesting to me about this particular environmental disaster is that there is a human tragedy within the environmental one. We often think generally about human tragedy being caused by environmental degradation. In other words, environmental degradation comes first, and human suffering follows. In this particular accident in the Gulf, the two are at least partly coincidental. The same event damaged the local environment and caused loss of life. Additional human suffering is likely to follow, but in this instance, this accident was both a human and environmental tragedy from the very beginning. What concerns me in the coverage and discussion of this tragedy is the relative lack of attention to the 11 souls who died far away from their homes and families. It occurs to me that our modern view of environmental degradation puts the environment on a pedestal above even the blessing of human life. Perhaps it is my choice of media outlets that is informing me of these events, but I hear almost exclusively about the environmental impacts of this tragedy and very little about the 11 people who lost their lives. Perhaps the sad tragedy of human death is too common and simply not as newsworthy as the ongoing calamity from a major environmental accident.

Accidental happenings certainly are newsworthy, and unfortunately, accidents occur regularly. There are accidents on our highways and plane crashes in the skies. Things go wrong at hospitals, at nuclear reactors, and even spacecraft explode. Certainly these events are not desired nor welcomed. They are not the direct, "normal" result of the activity in question but rather an unfortunate statistical anomaly of a tragic combination of abnormal factors. Of course, at some point, either implicitly or explicitly, we have assumed the risk of accidental happenings. As much as we try to make our activities safe, we have not been able to reach the point where accidents cease to exist. In the case of activities with potential to impact the environment, we have accepted the risk of environmental degradation due to accident.

Environmental degradation from accidental happenings still comes as a surprise to many of us. This happens only because we tend to forget the bargain we make whenever we undertake grand actions that have great potential for environmental harm. The only real surprise should be the specific timing and severity of the accident itself. It may be unpredictable and unplanned, but it should not be unexpected. As imperfect human creatures we make mistakes from time to time in matters large and small. Sometimes our mistakes, or an unfortunate combination of human and natural factors, lead to undesired outcomes.

We should also expect environmental degradation to occur from many of our routine activities. We know with near certainty the amount of pollution that will result from burning coal to produce electricity at a power plant, or using gasoline to power our automobiles, or even stoking our fireplaces to warm ourselves in the winter. We suffer through the smog and noise of a daily commute on crowded urban highways fully expecting the unpleasant conditions. We send out fishing fleets to harvest the fish from the oceans fully knowing that at some point we may put some species in jeopardy. Environmental degradation is not "accidental" in these cases even if it is sometimes ignored.

So, then, if environmental degradation can range from expected to accidental and from annoying to tragic, how do we judge the validity or severity of a claim of actual or impending environmental harm? The answer to this question depends very much upon the individual or group making the judgment. From time to time expert panels have assembled to make this judgment and gauge the severity of the risk. The

Intergovernmental Panel on Climate Change (IPCC) is such a group, assembled in 1988 to tackle the issue of global warming (IPCC 2010). They are now at the center of this international debate in defining the severity of the issue and proposing appropriate measures to counteract the threat. The Club of Rome is another such group (Club of Rome 2010). It is a global think tank of longer standing but smaller in membership. It was formed in 1968 and published a now famous report in the early 1970s entitled *Limits to Growth* (Meadows et al. 1972). It predicted economic collapse due to a growing human population overwhelming the planet's capability to provide the goods and services they demand.

This book, however, is not about expert panels telling us what we should do. History has shown that expert panels are not always right. The world is not flat, and the sun does not revolve around the earth. It has been nearly 40 years since *The Limits to Growth* was unveiled, and we are still here with our economies still functioning. This book is about individuals making up their own minds about environmental dilemmas and deciding how to respond. Part of this process is to gauge the environmental issue itself, and part of it is to gauge the efficacy of the response, particularly with respect to other impacts that the response may bring. This first pillar of the metric, environmental degradation, is only one part of the story. We must put this pillar into proper perspective, and then we must use the other pillars as well to put the entire issue in proper perspective. More about the other pillars in the chapters to follow, but just how do we put environmental degradation into its proper perspective?

First, it is important to acknowledge that environmental impact is always with us. By our mere existence we change our environment. We remove oxygen from the atmosphere and replace it with carbon dioxide, as do all the animals that share our world with us. We rob the world of relatively pure and pristine water and return water contaminated with wastes. We steal fruit from the trees and tear open the ground to sow seeds to grow our grain and vegetables. We slaughter animals to fill our stomachs, cover our skin, and adorn our bodies. We extract minerals and metals to make things. We harness the energy from within the earth to make us more comfortable and more prosperous. Each of these activities has an impact on the earth. Some impacts individually are quite small and perhaps even imperceptible, but they do exist. Literally, we cannot survive without impacting the environment.

We may think of our environment as the natural world around us, or we may think of it as simply the world around us encompassing both nature and humanity's improvements of buildings, roads, and all sorts of machines. If we accept for a moment the broader definition, the environment is the fields, forests, lakes, rivers, farms, and cities that we experience. We have grown comfortable with the human–nature dichotomy that we experience every day, and that seems to be ever present in the way we view the environment. We see a hazy midafternoon sky and notice the industrial smokestacks protruding into air. We hear the pounding ocean surf and notice the jetsam left behind by the high tide bearing the indelible human mark of discarded trash. We are presented the picture on the news of a burned-out section of jungle in Brazil, and we might think of a poor farmer struggling to earn a living and feed his family. It has become almost impossible for many of us to view the natural world without also seeing our human encroachment upon it. We exist within the environment, and it exists within our being and in our minds. Whether or not we consciously acknowledge it, we are linked.

It is common to observe our environment and notice examples of environmental degradation all around us. Its impact can range from strong to subtle and logical to emotional. Pollution is probably the most common and obvious example, but changes to the landscape or the abundance or scarcity of particular species are noticeable as well. Some environmental issues are spread over large areas or large spans of time, such as global climate change. Some exist within judgments of aesthetics, ethics, and societal values, such as species extinctions or the construction of large dams and reservoirs. The detriment to humanity can be difficult to ascertain and measure. Many of the environmental issues that we hear about may be very indirect and may not personally impact us in any meaningful way.

In its simplest terms, environmental degradation is a negative environmental impact. That then begs the question of what is a positive environmental impact. After all, by our mere existence, we humans exert a negative environmental impact upon the earth, right? So now we have a couple of fundamental questions to address. Number one, since our basic human needs and functions create negative environmental impacts, what really is the difference between environmental degradation and environmental impact? And number two, how do we reconcile a positive environmental impact against this backdrop of humanity creating negative environmental impacts by our very nature?

With regard to the first question, the difference between environmental impact and environmental degradation is mainly one of degree. Environmental impact becomes environmental degradation when a noticeably significant negative effect occurs, one that alters the balance of nature. While humanity exerts an environmental impact upon the earth, it has historically been balanced. While we humans consume oxygen and produce carbon dioxide through respiration, plants do the opposite and, theoretically at least, the atmospheric concentrations of oxygen and carbon dioxide stay in relative balance within a meaningful human timescale. While we harvest and consume plants and animals, nature tends to replenish itself, and the population of these species remains balanced. If our impact upon the world becomes unbalanced, then it is appropriate to use the term environmental degradation. Unbalanced, in this sense, refers to a significant change in nature's composition or function. If we hunt the great whales to extinction, then we have degraded our environment, for part of the ocean ecosystem will not behave as it once did. If, as the global warming theory proposes, humanity changes the composition of the atmosphere, resulting in extreme temperatures, large changes in sea levels, and increased incidence of violent storms, then we have degraded the environment, for we have altered the way the atmosphere functions. If we dam a river or drain a lake, we have degraded the environment, or at least the local environment, for we have changed the landscape enough so that it will not function as it did previously.

Our second question involved the difference between positive and negative environmental impact. This is really a historical question as past environmental degradation is what allows present beneficial environmental impact. Positive environmental impact is repentance for past sins. It is righting the environmental wrongs we thrust upon our environment years or decades ago. As such it is somewhat of a relative measure. For example, we may consider solar energy to have a positive environmental impact, the reason being that it produces electricity without creating any air pollution. In reality, and with respect to air pollution, we only see this as positive because it seems to be a better alternative than burning fossil fuel to create electricity. Nature is no better off because we manufacture solar panels and convert sunlight into electricity. Nature benefits because we could have done far worse.

As we will discuss more fully later in this chapter, practical environmentalism requires a judgment to be made about environmental degradation. This judgment is also a matter of degree in that not only will we differentiate between positive and negative environmental impact but we will also differentiate between large and small impacts, both positive and negative.

Certainly environmental degradation is important, but it is probably overly simplistic to think that environmental degradation is the same thing as environmentalism, or that it is the only basis for environmentalism. While environmental degradation is listed as the first pillar of practical environmentalism, I will later argue that it is technically no more important than the other three pillars. Technical considerations, however, do not really drive our modern environmentalism; emotion usually does. And so, it is admittedly unrealistic in our current view of environmentalism to think that the other three pillars of resource conservation, economic progress, and personal benefit are on an equal footing with environmental degradation. That is one of the main reasons why practical environmentalism, as defined here, is a new, interesting, and hopefully beneficial way to approach environmental issues. Environmental degradation is still critically important, but not solely important.

Our emotions and our reactions to the perceived degradation of nature is the fundamental driving force of our modern version of environmentalism. Our experience of an injured earth can have a profound impact upon us individually and collectively. The sight of a collection of tree stumps, the remnants of an old-growth forest clear-cut to feed the timber or paper mill is a powerful image that can stir powerful emotions. The stench from a putrid collection of garbage, cast off from our modern consumerism and headed to a landfill somewhere in the countryside, can trigger new ways to see that which we discard in the trash can under the kitchen sink. The roar of jets taking off from our many airports or the din of rush-hour traffic plodding along our many highways can sound an alarm in our minds that perhaps if our environment was better, our life could be better, too. Perhaps these examples are overly dramatic, but I hope it makes the point of how easy it is for environmental degradation to connect with powerful emotions within us.

The despoilment of the earth, in all its various forms, is what all the drama and excitement is about. According to many of our modern media outlets, this is where the danger lurks. This is the proposed menace to our future, and this is certainly logical to a degree. It makes a great deal of sense to wonder just how long we can continue to pollute our environment before it becomes unbearable. How much trash can the landfills hold? How much pollution can the sky and sea stand? How much warmer can our climate become before we are really in hot water? These are tough questions to answer, maybe even impossible questions to answer with complete certainty.

These types of questions fall within the "How Much Is Too Much" category. They ask if our human actions are overtaxing the environment and its ability to function in a way that will ultimately diminish the planet's capability to support humanity and the other species that share this globe with us. These types of questions are notoriously hard to answer. Many of the confounding factors mentioned in the previous chapter are at work with these types of questions as well. Paranoia, the problems with percentages, and our reliance on prediction to quantify environmental threats can easily find a home within the "How Much Is Too Much" category of questions. Yet these types of questions are undoubtedly a strong component of our modern environmentalism and really are necessary at a theoretical or pedagogical level at least.

Environmentalism should be able to address these questions in theory even if it cannot answer them in practice. These questions are essentially of a philosophical nature and deal more with the theoretical structure of our own particular brand of environmentalism rather than a point of fact. They test the logic of perspective-based environmentalism. If you recall from Chapter 3, there are many different ways we may choose to view the environment, and these different perspectives would likely result in very different answers to the "How Much Is Too Much" type questions. Cornucopians might say that there is no such thing as too much trash. Trash to them might be a resource that humanity will undoubtedly find a means to profitably use one day. Doomsdayers, on the other hand, might say that we already have too much trash and that the landfills that contain these enormous mounds of festering filth already threaten our local water supplies and will one day poison us through our own kitchen faucets. Someone who holds the Nonbeliever perspective or someone with the Spaceship Earth view would probably come up with other responses to the question. So the questions of how much trash is too much, or how much air pollution can we stand to breathe, will likely produce multiple answers based on the numerous and varied environmental ethics that exist. In effect, these types of questions are very helpful at describing and illuminating particular environmental perspectives. They are just not very helpful in the practical level, and it is at the practical level where we actually consider doing something to lessen environmental degradation.

Practical environmentalism asks a slightly different question about environmental degradation. It simply asks if things are getting better or worse. It does not deny the existence of the highly academic, tough-to-answer questions mentioned previously. It just does not find these types of questions to be very useful in a practical sense. Practical environmentalism asks questions such as

- Does the construction of the power plant within my town have a negative impact to the environment?
- If I drive an electric vehicle, will that provide a benefit to the environment?
- Will banning the production and sale of incandescent light bulbs benefit the environment?
- Does no-till agriculture provide environmental benefit?

Practical environmentalism addresses environmental degradation in a very specific way. It focuses on an action, activity or decision, and then asks you or me to judge the environmental benefit or detriment associated with that particular activity or decision. Practical environmentalism desires to separate the evaluation of environmental impact from all our other biases, considerations, judgments, and perspectives just long enough so that we can consider the environmental impact with a measure of mental liberty and clarity of thought. As mentioned previously, Practical environmentalism is a deconstructionist model and attempts to strip away extraneous and unnecessary information. For many environmental issues we need to create enough room within our minds to ask ourselves enough questions to reasonably inform ourselves of the likely environmental impacts of the action we consider. And oftentimes it would not hurt to do a little research in order to obtain the additional information needed to produce a reasonable and rational opinion.

Practical environmentalism also enables us to separate the impact of our action with the impacts of other actions. This allows us to reflect whether our action, even if it produces a negative environmental impact, is a significant contributor to environmental degradation. This is a critical distinction when we consider that some types of environmental degradation have both human and natural causes. For example, volcanic eruptions can release a large amount of air pollutants into the atmosphere. Significant emissions of particulate matter, sulfur dioxide (SO_2), hydrogen sulfide (H_2S), carbon dioxide (CO_2), hydrogen chloride (HCL), and hydrogen fluoride (HF) are typical. These compounds are also released when burning coal or heavy oil or even biomass in some cases. If were we contemplating action to drastically reduce sulfur dioxide emissions from power plants worldwide, it would certainly be reasonable to compare the anthropogenic (an academic term that I learned in graduate school, which means caused by humanity) emissions of sulfur dioxide compared with the natural emissions. We might judge the environmental impact of this action very differently, depending upon the relative contributions of the two sources. If we learned that the average volcanic emissions of sulfur dioxide were 100 times greater than power plant emissions, we might decide that the environmental impact of the power plant emissions was already very slight. If, on the other hand, the ratios were reversed and anthropogenic emissions were 100 times greater than natural emissions, then the environmental impact of our proposed action could be very great. Obviously, a little research to determine the relationship between the two types of sources in this example could enlighten us quite a bit. Practical environmentalism expects us to invest enough time to research these types of questions. Practical environmentalism asks us to be reasonably well informed before we pass judgment.

This ability to objectively separate the impact of our action from the impact of other actions is also important strictly on a human scale. If we have a plan to reduce our water use, for example, will it have a very great impact if we are already very frugal with our water consumption and our neighbors are very wasteful? Would it matter if a single large factory in town used more water than all the residents combined? It could be reasonably argued that an initiative to save water at the factory would have a greater environmental impact than saving water at a particular residence.

These types of questions are filled with social implications, some of which we have already touched on in Chapter 3 on environmental ethics. These social implications occurred at the local scale in the examples cited previously, but could easily stretch to the national or international scale. On the worldwide stage of global climate change negotiations, nations bicker with one another over their carbon dioxide emissions relative to other countries' emissions levels. Should a small, less-developed nation be expected to commit to the same carbon reduction levels as a large, prosperous one, perhaps one that consumes a relatively large amount of energy in proportion to its population? Is not the environmental impact of the large nation's emissions much greater than the small nation's? Does it make any sense at all for the small nation to reduce carbon emissions if the large nation will not?

Practical environmentalism allows us to incorporate these nuances of environmental degradation within our decision-making efforts. It uses a relatively simple scoring system, one that can be used whether the issue is big or small and whether we have lots of data or precious little good information. The system itself is common to all four pillars and easy to use. It allows both beneficial and detrimental judgments to be made within the same system.

The scoring system employed translates our judgment of environmental degradation into a numerical scale of five numbers: −2, −1, 0, 1, and 2. Negative values, as you might guess, imply a detriment to the environment, positive numbers a benefit. Zero represents a neutral impact to the environment or one so small or so uncertain that it is unreasonable to qualify it positively or negatively. The number "2" represents a greater impact, either positive or negative, than the value of "1" and allows us a bit of latitude in ranking risks and rewards. Obvious environmental degradation with severe health impacts to humans would rate a −2. Environmental impacts of a less serious nature, or where predicted impacts are very controversial, might rate a −1 or possibly a 0. Environmental benefit would follow the same ranking logic with obvious or certain benefits scoring higher than more subtle benefits.

The system was designed specifically to have negative, zero, and positive selections. This is to make it easier to deal with more qualitative issues where quantification becomes difficult. A negative selection is clearly distinguished from a positive one, and we do not have to quibble so much as if we were trying to rate responses on a 1-to-10 scale. The negative/positive choice also allows us to easily combine results from the first pillar with the other three. We will describe this combination in more detail in later chapters, but suffice it to say for the moment that, if you have negative scores for each pillar, you probably should rethink your action.

Admittedly, this is a very simple system, and it might not work perfectly for all issues. It will work for many situations though, and its purposeful simplicity will make it easy to use. You will not have to consult an expert to tell you what to do. You will have to invest a little time, effort, and thought into informing yourself of the issue at hand and the likely effects of any proposed actions you plan to take as a response to the issue.

And here is one of the main reasons for writing this book. Are we not encouraged and expected to do something to prevent the environmental calamities that await us? Is not that the point of all the horrific predictions, the international uproar, and the programmatic reeducation of our environmental knowledge and ethics? We need to DO something, right?

While I intentionally write a bit sarcastically here to highlight my distaste for some of the current portrayals of the looming environmental catastrophes, I do agree that we should do something. We can and should do our part to make our environment better. We should consider changing our lifestyles and making different choices than we have in the past. The scoring system and the four pillars are meant to guide us to informed inclusive decisions when we attempt to do things differently for the benefit of our environment. Let us make a reasonable effort to ensure that changes we make for the sake of the environment do not backfire and cause us distress in other facets of our lives. Let us properly consider all four pillars before we start to doctor the environment. Let us "do no harm."

REFERENCES

Intergovernmental Panel on Climate Change (IPCC). 2010. http://www.ipcc.ch/ (accessed October 1, 2010).

Club of Rome. 2010. http://www.clubofrome.org/eng/home/ (accessed October 1, 2010).

Meadows, D.H., Meadows, D.L., Randers, J., and Behrens, W.W. III. 1972. *The Limits to Growth.* New York: Universe Books.

6 Resource Conservation— The Second Pillar

The first pillar, environmental degradation, represents what humanity does to the natural world. This second pillar, resource conservation, represents what humanity takes from the natural world. Nature is both a source of material wealth to humanity as well as a sewer for our wastes. In this chapter we will focus on what nature gives us.

Resource conservation does not make the headlines as often as environmental degradation. When it does, it is often related to a shortage of something we need or a sharp rise in price of some basic commodity. We typically pay attention to resource conservation during these times of crisis. In the early 1970s, for example, the United States endured an oil embargo from several Middle Eastern oil-producing nations who were displeased with US politics in the region. Suddenly, there were long lines at gas stations and prices were climbing fast. At the time, many Americans preferred to drive the large and powerful autos produced by the Detroit car companies. This was the era of the muscle car in America. Mustangs, Camaros, GTOs, and Chargers prowled the streets with loud, rumbling engines and enough power to spin their rear tires at will. Thirsty V8 engines were the norm, but were now at a relative disadvantage. Fuel economy was not their strong suit, and now their frequent refueling stops were more costly and often meant waiting in line with a lot of other irritated motorists.

Practically overnight, fuel economy became important. The American auto industry was not well positioned to respond to this change in fuel availability and cost. For decades the natural evolution of the American car was toward bigger and more powerful engines. With the cost of gasoline low, this was a very easy sell for corporate marketers to make. Engine size and horsepower were strong selling features. The automobile of the muscle car era was not intended simply to transport you from point A to point B. It was supposed to make the transit a bit of a thrill. The right car promised to add some entertainment to one's previously mundane daily travels. As an added benefit, it just might impress your neighbors or even a potential girlfriend.

But times changed, and conserving gasoline, just one of nature's many resources, suddenly became important. In this case conservation did not become important because nature was just about tapped out of the oil we needed to make gasoline, but rather we in the United States could not get as much as we wanted. One single resource manipulated to be in short supply caused economic and social havoc within one of the most prosperous and powerful nations on earth. Arguably, the turmoil eased over time, but this is a good case study of just how much we can depend upon natural resources, and how disruptive it can be when they are not available in ample supply.

We consume nature's resources constantly, and it is not merely one or two that we require. With every breath we take and with every sip and morsel that crosses our lips we depend upon nature. Natural resources sustain us minute by minute, hour by hour, day by day, and year after year. Our possessions, including clothing, homes, automobiles, toys, and books, can all be traced back to Mother Earth in some form or fashion.

Fresh clean air is a resource. So is drinkable water and fertile soil. Forests, deserts, mountains, rivers, lakes, oceans, mines, and sunshine are all resources. Money is a convenience of our imagination. A lump of coal is real.

I think of resource conservation as the forgotten sibling of environmental degradation. They are strongly linked, however, in that they both bear directly on the environment's capability to provide for humanity. Where environmental degradation often deals with the demands we put on the natural world, resource conservation concerns itself with the supply side of the equation. We literally draw our sustenance from nature's resources. Regrettably, it seems that we have en masse forgotten this truth. Food comes from restaurants and grocers, clothes from the mall, and heat and electricity are merely automatic payments drawn from our bank accounts or credit cards on a monthly basis. How fortunate we are to be able to live so easily with this delusion!

I am in no way advocating a return to a simpler, more basic existence. If I had to grow or hunt the food for myself and my family, I am afraid we would be a very skinny bunch. I am, however, advocating that resource conservation be an integral part of our environmental decision-making process. I was tempted to list resource conservation as the first pillar of the metric, but I believe that it has been a distant second for so long in most of our minds that I decided to keep it second in this discussion as well. Resource conservation is a topic, however, that is near and dear to my heart, admittedly a bias on my part. I have always been intrigued by the notion of self-sufficiency with respect to our basic needs of food, shelter, and energy, and resource conservation plays a big part within that framework of thought.

In terms of the resources that we draw upon for our survival, some may argue that the ultimate resource is the human intellect, and I would not argue too strongly against the assertion. However, humanity has to have something to work with. Basic needs must be fulfilled. We require a minimum caloric intake, a minimum level of hydration, and the ability to regulate our body temperature to approximately 98.6 degrees Fahrenheit (37°C). Luckily for many of us, these pesky realities are things learned in elementary school and then largely disregarded as trivial and unimportant. For most reading this book, the challenge is to limit caloric intake to a reasonable amount, not to search for enough food to continue existence. We choose whether to drink bottled water or risk swallowing water from our faucets at home. We buy our clothing mainly for appearance and often for the name on the label or the insignia stitched into the cloth. Some people, typically those who are much closer to my children's generation than to mine, go so far as to purchase clothes with holes, tears, and rips intentionally added. These imperfections have become a symbol of style instead of a sign of age and overuse. For many around the world these choices represent great luxuries. For them, sufficient food, clean water, and appropriate attire may be truly scarce.

For many of us, our lifestyles have advanced to the point where basic needs are taken for granted, luxuries have become necessities, and our desires know few bounds. This advancement has occurred precisely because of the creativity and imagination of the human intellect coupled with a desire for wealth and security.

This is our human nature. We are blessed with a desire to improve ourselves and our situation. We strive to give our children opportunities that their grandparents never had. Wealth is security and it transforms us. We no longer think much about the basics of survival. Instead we focus on our investments, the grandeur of our homes, and where we will vacation next month.

Again, allow me to repeat that I am not advocating deevolution of our economic progress. I worry about my investments, too. I wonder if I will be able to finance a college education for my children and save enough for my own retirement. While my home is not particularly grand, it is comfortable, and I am proud and happy to occupy it. It provides security for my family. It is a good thing.

What I am advocating is the specific inclusion and strong consideration of resource conservation into environmental decision making. This is critically important for two main reasons. The first is that the transformation of natural resources into consumable commodities has an impact on the environment, usually a negative impact. The wanton and reckless consumption of natural resources often leads directly to environmental degradation. For example, overfishing, overfarming, overtimbering, overmining, over-grazing, and overpumping aquifers can lead to a degraded and less productive environment, at least at the local scale. You do not have to search too hard or too long to find examples of these losses to the balance sheet of natural resources that sustain us.

The second reason is that resource consumption is the basis for our economic systems now and certainly into the foreseeable future. It is truly the lifeblood of our societies. Just as overexploitation of natural resources can lead to environmental degradation, it can also lead to economic and social degradation. Overfishing, for example, not only depletes the stock of fish and puts marine species in jeopardy but it also depletes the economic and social stock of the towns and villages that depend upon the fishing industry. When fishermen have trouble catching enough fish to earn a decent living, the businesses that serve them also feel the impact of slow sales and decreased income. As a local economy falters, people are forced to look else-where for employment. Many people move away in search of better opportunities for themselves and their families. Housing markets may suffer. School enrollment may shrink, and a town, even an entire region, can be stressed and changed.

Perhaps an accounting analogy is an appropriate way to highlight the importance of resource conservation. Think of our natural resources as an interest-bearing checking account with a relatively large beginning balance given to us by Mother Nature herself. Left utterly alone, the wealth of the account grows steadily albeit slowly. But we cannot leave it alone. We must draw from the account periodically. We must write the checks that sustain our survival. What is important is to compare the value of the checks with the value of the interest. Once we begin writing checks that are greater than the naturally occurring interest, we begin to deplete the account.

Mind you, this interest-bearing checking account of ours is very large. The planet is very wealthy indeed. We have been writing checks against our account since before the dawn of recorded history. These checks have typically been minutely small and limited by a relatively tiny human populace and infantile levels of technology. Things sure have changed though. Humanity has grown up. We are far more numerous and far hungrier than we have ever been in the past. The checks we write are bigger than ever before.

Centuries ago, we were forced to live off the interest of nature. We were not smart enough to know any different. We struggled to produce enough food for ourselves, and could not begin to imagine marketing perishables to people half a world away. Energy needs were met largely from wood, and wood was produced by Mother Earth every year. Transportation came from animals, horses, camels, and elephants, or on foot. Mother Earth took care of the animals, too. Our clothing was made from cotton, wool, or hides, which was perpetually available with a minimum of fuss. Most of the goods and services we consumed came from a benevolent Mother Earth who dispensed these essentials to humanity from the interest of nature.

We are much brighter now. We have learned to harness and transform the wealth of nature to make our lives much easier and more pleasant. We have tamed the forces of nature, and they are at our beck and call. Electricity no longer exists solely in the huge burst of energy within a lightning bolt. It now lights our homes, powers our factories, and plays music from all sorts of tiny devices. The recreation of nature is no longer such a mystery, and through genetic engineering, we have begun to tune it to our preferences. We have split the atom and unleashed the vast amount of energy within it. We have subdued the pull of gravity and soared into the air to travel across continents quicker than we used to be able to cross a state. A few of us have traveled into space, and a very few have even placed their feet on the surface of a globe not our own. We have done all these things by combining the power of the human intellect and learning with the vast wealth of our natural resources. We have figured out how to dip into the balance of the checking account. We have graduated into a world where we can write very big checks indeed. We are no longer limited to the interest alone.

This new capability must be tempered with cautious insight into how nature works. We must now be keenly aware of how big our checks are. We have also come to realize that this simple analogy of a single interest-bearing checking account is far too simple. A superior analogy would be to consider that there are actually many accounts that earn interest at different rates. Our actions, the fictitious checks in this analogy, will impact some of nature's accounts more than others. While we have drawn down some accounts very severely, others are still rather robust. Some, of course, have been closed, for example, when a species goes extinct or a lake dries up and disappears.

Let us now delve a bit deeper into some specific characteristics of nature's bank of resources and begin to highlight the different type of accounts that are present within the natural world. Simply put, resources are things we need to survive and prosper. In the context of this book, resources are commodities made available to us through the bounty of nature. The word *resources* here is used synonymously with the term *natural resources*. If you allow me to disassociate humanity and nature just for a moment, we may consider that some resources are not natural. Time and money are examples of resources that do not come from the earth. These are human resources really, and while they are very critical in most aspects of our daily lives, they do not qualify as natural resources per se. The human intellect is another example of a nonnatural resource emanating from humanity itself rather than strictly from nature.

It is also instructive to differentiate between the types of natural resources. Three main categories are often used to describe natural resources. These categories are perpetual, renewable, and nonrenewable. They differ with respect to availability and origin, where origin refers to nature's process of creating the resource. In our modern way of thinking, we often do not differentiate between perpetual and renewable resources and lump them together at times. I think that it is instructive to consider them separately, however, and so we will differentiate them here.

Perpetual resources are continuously available and not within human capability to diminish in any significant way. The energy from the sun is a good example of a perpetual resource. We humans have absolutely no control over the energy liberated from the surface of the sun, nor the amount of solar radiation that falls upon the earth's surface. The energy emanating from the sun is expected to last for billions of years and will not change appreciably over time, as humanity understands the concept of time. Notice though that I did not use the term "sunlight" here as an apt example because sunlight is a very local phenomenon. It depends upon cloud cover, tree cover, and even the height and orientation of nearby buildings. We humans do have the capacity to influence the amount of sunlight that strikes a particular piece of ground, so on a local scale sunlight does not fit our definition of a perpetual resource.

Air and water also fit the definition of a perpetual resource if we consider these resources without regard to quality. Air exists within our atmosphere, and we have little control of the quantity that exists. We certainly can influence its composition, mainly by contaminating it with added pollutants, but we cannot readily change its volume. Likewise, the total volume of water on the earth is generally constant. Humanity can pollute our water resources just as we can pollute the air. We can also have some influence over the location of water, be it in underground aquifers or freshwater lakes or in saline oceans, but we have relatively little impact on the total volume of water that exists on the planet.

Another way to view perpetual resources, particularly those that come from the earth, is the ability to recycle the resource. As mentioned, air and water are perpetual resources largely because they are recycled through natural processes. Some minerals and metals may also be considered perpetual resources. We extract aluminum ore from mines, for example, and use some of it to manufacture cans and containers for our favorite beverages. We can recycle the aluminum cans to make more cans for more beverages. The aluminum itself is not destroyed in this process. We merely concentrate it from ore to metal to make it more useful. A basic characteristic of a perpetual resource is humanity's inability to change the resource to such an extent that it is no longer useful in a timescale appropriate for humanity.

Renewable resources, on the other hand, are subject to humanity's management and, unfortunately, mismanagement sometimes, too. Renewable resources are reborn frequently and, with proper management, can last indefinitely. Animals are a renewable resource, literally reborn every so often as long as their population remains healthy and large enough to be viable. If placed under undue stress however, their numbers could shrink to the point of extinction, to the point of no return. Selfishly, I would hate to see cows and chickens become extinct. I have gotten very accustomed to the products they provide us. I would miss the eggs on my breakfast plate and the butter on my toast.

We humans do have the capability to drive a species to extinction, and we have prove it. The dodo bird of the Indian Ocean of Mauritius is a prime example

(Bagheera 2010). Through hunting and the human introduction of competing species, the dodo could not survive, and faded into history as a species. Unfortunately, once gone, they are really gone forever. We humans have not developed the ability to re-create an extinct animal species.

Pure, clean drinking water is another renewable resource. The adjectives pure, clean, and drinking are necessary to differentiate it from the perpetual resource of water itself. We can mismanage our water supply to the extent that it becomes unusable as a source of drinking water. We can contaminate it with harmful pesticides or radioactive material such that we dare not drink it. We can pump so much water out of an underground aquifer that it becomes exhausted, or so high in mineral content that it becomes unusable. The cities and towns that depend upon such an aquifer for their drinking water would be faced with a serious problem.

We often think of solar energy, wind power, and hydroelectricity as renewable forms of energy. I would consider solar energy and wind power as perpetual resources, but I would not argue too strenuously because perpetual and renewable resources have much in common. The key distinction is that renewable resources can be depleted if managed improperly. We really cannot do much to stop the sun from shining, but we can pollute the air so drastically that the intensity of the sunlight is reduced before reaching the solar collectors installed on my roof. I really do not know how we could stop the wind from blowing or the rain from falling upstream of our hydroelectric dams, though. Humanity can definitely impact many renewable resources, however. Forests can be cut down, topsoil can wash away, and species can become extinct as a result of human action.

Trees are a prime example of a renewable resource. They grow a little bit every year. Their trunks become larger, and their top branches reach higher and higher into the sky. Some produce fruit or nuts or even syrup on an annual basis. We can harvest trees for their wood, and, by managing the rate of harvest as well as the rate of replanting, we can count on a reliable supply of wood year after year without depleting the overall stock of trees on the planet. If, however, we cut them down too fast, we can deplete the stock of trees and diminish the resource. A newly planted sapling is far different from a mature apple tree loaded with fruit, or from a majestic redwood large enough for a car to drive through at its base. A relatively long time lag exists between when we plant a tree and when it begins to produce resources in significant quantity. Trees are renewable as long as the growth rate of existing trees plus the planting rate of new trees equals or exceeds the harvest rate. Figure 6.1 is meant to remind us of the beautiful complexity of trees.

I could be accused of undue simplicity with this example of trees. Some might point out that I neglected to appropriately consider the biological and ecological differences between old-growth forests and tree plantations. Others might say that there is a huge environmental difference between clear-cutting and selective logging. You might remind me of the potential harm to species from habitat destruction or the potential disruption to the hydrologic cycle by disturbing the landscape. All of these points could in fact be pertinent in one particular case or another, but some of these points tend to blur the line between the resource conservation and environmental degradation pillars. The simplicity of my example of trees is purposeful in reinforcing the concept that the resource conservation pillar attempts to focus solely on

FIGURE 6.1   The simple yet wonderful tree. (Illustrated by Mark Benesh. With permission.)

natural resource implications. The environmental degradation pillar should be used to account for specific environmental impacts. Practical environmentalism relies on each pillar to adequately represent our concerns with each environmental action that we consider. We can then combine the results of each pillar into a holistic measure to facilitate our decision making. There will be more on this concept in the chapters that follow.

Nonrenewable resources are not continuously reborn and can essentially be thought of as fixed in supply. A basic characteristic of nonrenewable resources is that using them changes their form so drastically that they cannot be made anew. This characteristic often differentiates nonrenewable resources from perpetual resources. Some resources of both types exhibit the characteristic of being fixed in supply. There is a certain amount of water on the face of the earth, and there is a certain amount of coal buried within it. Water is perpetual primarily because we cannot destroy it. We can pollute it and render it useless for a given purpose, but this does not destroy the water itself. The two hydrogen atoms bonded with the single oxygen atom continues to exist. Even if we decompose the water molecule into its basic components of hydrogen and oxygen, nature will recombine them at her first opportunity, and the volume of water on our planet will not have changed.

Coal, on the other hand, is usually changed by our use of it. We typically "burn" coal to liberate energy. Where we used to burn it to heat our homes and buildings, or drive a locomotive engine, we now burn it to produce steam that turns a turbine to make electrical power. In fact, approximately half of the electricity consumed in America comes from burning coal (EIA 2010). The carbon and hydrogen atoms in the coal combine with oxygen in the air to produce carbon dioxide and water and significant amounts of energy. The coal is destroyed as its chemical structure is transformed into more fundamental components. It will require millennia for nature to recombine carbon and hydrogen into the form we today recognize as coal.

Re-creating nonrenewable resources is practically impossible due to either the very long time required or the great amount of energy required to reform them. Even though nature created them in the first place, it took such a great length of time that the rate of creation is essentially and practically zero. Nonrenewable resources are sometimes called fossil resources, alluding to their creation eons ago. This is particularly true with respect to fuels. Coal, oil, and natural gas are often termed "fossil fuels" and are considered to be nonrenewable. Some people use the term "fossil water" when they speak of water in deep underground aquifers. They use the term to stress that this underground water came from rainfall thousands or millions of years ago that eventually filtered its way through the earth to where it now lays contained. If we were to pump the water out of this aquifer, it would not be replenished within our human timescale, and so locally it could be considered a nonrenewable resource.

Another resource that nature provides, and one that most of us have very little understanding of, is nuclear energy. The energy within the atom itself is a very mysterious commodity with only a relatively few well-studied and knowledgeable individuals having a basic understanding of its inner workings. The majority of us understand nuclear energy in only the most basic terms with reference to nuclear bombs, nuclear power, or nuclear radiation. Within the past century, humanity has fully entered into the nuclear age, and we have unleashed incredible amounts of power as well as an incredible potential for harm to the environment and to ourselves. The huge amount of energy contained within the tiny atom inspires awe as well as fear.

Nuclear energy is a significant natural resource. It satisfies approximately 13.5% of the world's electricity demand (IEA 2010), but in so doing, it creates waste products with the capability to seriously degrade the environment. Nuclear energy can also be utilized as a weapon with devastating and long-lasting results. The two atomic bombs that effectively ended World War II are relatively impotent compared to the nuclear arsenal that exists today in many countries all over the world. It is a complex and complicated subject from scientific, social, and environmental perspectives. Within this chapter, we will deal with it very simply in terms of how it fits within our view of natural resources. In later chapters we will discuss it more comprehensively as an environmental issue subject to the practical environmentalism framework and approach.

For our purposes here, nuclear energy is a nonrenewable resource based on the presumption that the fission of uranium or plutonium is the main source of nuclear power. Both of these minerals are present in the natural environment as relatively dilute ores. They are mined, concentrated, and processed into fuel, which is fed to nuclear reactors to produce power. The controlled fission of uranium is the basis for the nuclear power industry in the United States, but plutonium is used in some reactors outside of the country. Fission is a reaction in which the relatively large nucleus of the uranium or plutonium atom is split into other smaller atoms, typically isotopes of iodine, cesium, strontium, xenon, and barium, for the case of uranium fission (World Nuclear Association 2010). This reaction is exothermic, meaning that it produces heat. The heat generated within the reactor can be transferred to a working fluid, often water, to produce steam, which in turn drives a turbine generator to produce electricity. Similar to the combustion reaction for fossil fuels where hydrocarbons are consumed and other compounds produced, the uranium within a nuclear reactor is also consumed, being fundamentally changed during the fission reaction into other chemical elements.

We will purposefully neglect the concept of nuclear fusion as a natural resource. Nuclear fusion is the reaction that occurs within the sun and stars. Nuclear fusion within our sun is an atomic reaction where hydrogen atoms are fused together to produce helium atoms and a great release of energy. We humans have developed a basic understanding of this reaction and have even been able to replicate it at very small scale in a laboratory environment for very brief periods of time. We have not, however, been able to produce a system that will convert hydrogen to helium in a self-sustaining manner with the production of excess heat to power a turbine and generate electricity. We have not yet proved ourselves capable of harnessing the potential of atomic fusion to power our world. As such, I will leave the discussion of the raw material for nuclear fusion, typically hydrogen as a natural resource, to future generations.

Another category of nonrenewable resources is mineral ores. We have just discussed the radioactive minerals uranium and plutonium, but, of course, many other minerals exist. Mineral ores are formed by geologic processes on geologic time scales. Therefore we typically do not think of mineral ores as being reborn within a timescale that is meaningful for our thinking about using resources.

You will find that, within the broad category of minerals, some can be thought of as renewable and some as nonrenewable. The ore that we mine is usually thought of as nonrenewable, but the mineral we extract may be considered at least partially renewable via recycling. Consider the example of precious metals. While they meet one of the nonrenewable criteria in that they are not continuously reborn, they fail the other criteria in that their use does not change them so drastically that they cannot be used again. We may encounter them in nature as ores, but we often process these ores into their purer metal form. We might manufacture something using gold or platinum or silver, but when we are done, we can often recover and recycle the precious metal within the product. In fact, we already recycle many products specifically for their metal content.

There are many minerals that we extract from ores, and there are many uses for minerals within our industries. Some are used as a raw material to make a product, while others are used as an additive or reactant to drive a desired chemical process. In general, if our usage of the mineral transforms it such that it cannot be easily recycled or reused, then we should consider it as nonrenewable. If, on the other hand, we extract a mineral from ore that is recycled very easily, we should consider categorizing it as a renewable resource.

So, most often, our consideration of nonrenewable resources revolves around fossil fuels and nuclear energy. While some minerals might be considered nonrenewable as well, their scarcity or their potential for environmental degradation has not risen to a high-enough level to capture our attention. Let us discuss the most common category of nonrenewable resources, fossil fuels, a little more deeply.

We theorize that fossil fuels were created millennia ago by the decomposition and compression of biological matter and have been buried underground ever since. It was only in our recent past that humans figured out how to extract these resources from the ground and learned how to employ them for our benefit. We dug or drilled deep into the earth to reach the fuel and to eventually liberate this ancient energy and transform it into heat for our buildings, fuel for our cars, and electricity for our modern economy. We began to deplete the massive stock of fossil fuels that Mother Nature had stored for us over the centuries.

Remember that the critical characteristic of fossil fuels, and nonrenewable resources in general, is that once used, they are gone forever. How then do we account for their use? How do we justify using them if they can never be replaced? This is an economic as well as a moral question. The economic question relates to the value of using nonrenewable resources now versus the value in delaying their consumption, either for our own use in the future or possibly for use by our descendants. Does their immediate use today convey sufficient benefit to overrule their use next year? The moral question is concerned with who will reap the benefit of the use of the resource, ourselves in the present or those who might use it in future years.

The first step in answering these types of questions is to assess just how much of the resource exists and compare that to how much we use over some timescale, usually over a year's time. That calculation—let us call it a resource longevity calculation—will give us a rough estimate of how long the resource will last at current consumption rates and make it easier to identify the appropriate time period to assess future benefit and future beneficiaries. These longevity calculations are fairly well publicized for fuels. Organizations such as the Energy Information Administration (EIA), the International Energy Agency (IEA), and the World Resources Institute (WRI) regularly publish articles and reports covering usage rates and known reserves of fossil fuels. So, with a little research, you can determine how much oil is left in the ground and approximately when it will run out.

This type of information is fairly easy to gather but rather difficult to understand in highly accurate terms. One might surmise that, because this information is published by well-respected organizations, it is very accurate. The contrary proves to be true, however, in that published information about fossil fuel reserves are in reality estimates, educated guesses at best. These estimates will vary over time and between different sources. Sometimes new oil or gas fields are "discovered," or perhaps we find out that a field is not as large as we had hoped. It is not an easy task to know with great certainty how much of any particular substance exists under the ground. Figure 6.2 shows an example of this uncertainty in a trend of natural gas reserves in the United States over the last 85 years.

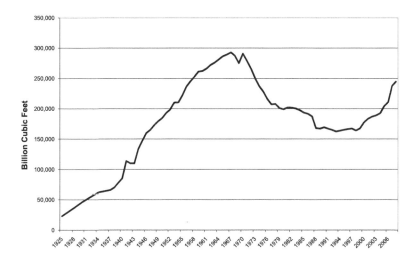

FIGURE 6.2   US natural gas reserve estimates from 1925 to 2008. (From Energy Information Administration [EIA]. 2010. http://www.eia.doe.gov/cneaf/electricity/epm/table1_1.html [accessed October 4, 2010].)

Figure 6.2 shows a pattern that we might not readily expect. Proven reserves of natural gas increased within the United States from 1925 through about 1968 and then fell until 2000. Since 2000 they have been increasing again. So how do natural gas reserves increase when we are constantly using natural gas, and nature is assumedly not creating any new gas? This variation occurs because of our lack of knowledge. Proven reserves of natural gas are fairly accurately known as their name implies. They are proved. Estimates of these reserves are based on known geologic and engineering data. Obviously though, we do not know what we do not know, and that can be significant at times. I would surmise that between 1925 and 1968 we were discovering new natural gas reservoirs at a rate that was significantly outpacing our usage of natural gas. The chart shows that US gas reserves increased approximately 11-fold during this period, growing from about 25,000 billion cubic feet (714 billion cubic meters) in 1925 to nearly 300,000 billion cubic feet (8,570 billion cubic meters) in 1968.

From 1968 to 2000, natural gas reserves dipped to about 160,000 billion cubic feet (4,570 billion cubic meters) and then rebounded to about 240,000 billion cubic feet (6,680 billion cubic meters). The discovery, or lack thereof, of new gas fields has a great influence on the behavior of this curve. Also relevant are changes in technology that allow more gas to be withdrawn from existing fields, essentially increasing the reserves. Figure 6.2 vividly illustrates the complexity of predicting just how much of any resource is available to us, and how confusing it can be to estimate how many years are left before any particular resource "runs out."

Another complexity that exists with resource reserve or longevity estimates is that we often do not know just how difficult it will be to extract any particular substance from under the ground. Remember that, in terms of nonrenewable fuels, we are after the energy content of the substance. If we have to expend more energy to retrieve a fossil fuel than exists within the buried fuel itself, then it is foolish and counterproductive to drill the well or dig the mine to get to it. In reality, economic constraints typically interfere before energy content constraints. Fossil fuels that are hard to reach are usually expensive to reach. It is costly to drill a well thousands of feet down, especially if the well is in an inhospitable location such as northern Alaska or in the middle of the Atlantic Ocean. Production and development costs, plus profit margins, need to be less than the market price of the fossil fuel in order for the extraction to be viable in purely economic terms. There can easily be political forces at play as well. The granting of drilling or mining rights may be subject to political preferences or nationalistic agendas. Production rates from existing fields or fuel pricing itself may be manipulated to achieve political gain. If it becomes too difficult for technical, economic, or political reasons, then the buried treasure is out of our reach and useless to us.

The other component of the longevity calculation is the usage rate. This is not a static figure either as the amount of fossil fuel that we use changes year by year and country by country for various reasons. Some countries tax fuel use at higher rates than others and thereby create an incentive to use less, either through substitution of another energy source or perhaps through more efficient processes that use less fuel to produce the same output. Some countries depend heavily upon mass transit or bicycles or walking as alternatives to automobiles and therefore have a relatively low per-capita usage rate of oil. Most countries are affected by general economic conditions and tend to use more fossil fuels when the economy is strong and less when the economy is suffering.

So, while the longevity calculation is well intended, it needs to be tempered with cautious realism. It would be nice to know with a high degree of certainty just how many more years of oil or gas supply we had. It would be nice to be able to manage an orderly transition from fossil fuels to renewable sources of energy. We could plan on using the fossil fuels right up to their point of exhaustion, all the while creating the technology and infrastructure necessary for the switch to the new fuel supply just in time. But our need for energy is neither orderly nor is it well managed. It fluctuates with economic conditions, social norms, and political realities. We are beholden to our current infrastructure and investments. We are subject to oil embargos and volatility in natural gas markets and occasional brownouts or blackouts of our electrical distribution systems.

Supply and demand of fossil fuels often struggles to find an acceptable balance. This is often evident as rapid increases or decreases in the price of some energy supplies. For example, Figures 6.3 and 6.4 show examples of the volatility of natural gas and crude oil prices over the last few decades. The natural gas chart represents prices for residential natural gas in the United States between 1981 and 2010. The crude oil chart represents the price of worldwide crude oil as a volume weighted average.

Obviously, the price of these two nonrenewable natural resources is not constant. Price frequently rises and falls. The natural gas chart shows this pattern to be fairly regular, while the crude oil chart appears to be more random. Noted on each chart are the dates associated with recent peaks and valleys. The summer of 2008 was clearly a bad time to be buying natural gas and crude oil. With respect to natural gas, the price rose nearly threefold from $7.3 per thousand cubic foot in December 2001 to $20.7 per cubic foot in July 2008. The price then fell in half to $10.31 per thousand

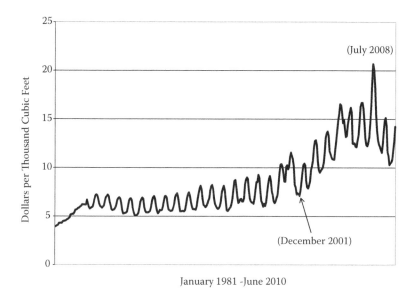

FIGURE 6.3 US residential natural gas prices from 1981 to 2010. (From Energy Information Administration [EIA]. 2010-2. U.S. Dry Natural Gas Proved Reserves. http://tonto.eia.doe. gov/dnav/ng/hist/rngr11nus_1a.htm [accessed September 15, 2010].)

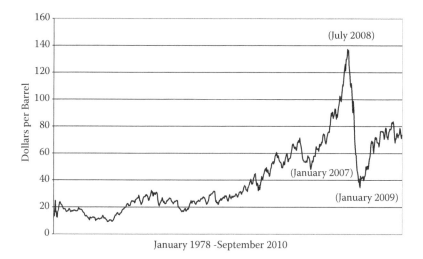

FIGURE 6.4 World crude oil prices from 1978 to 2010. (Energy Information Administration [EIA]. 2010-3. World Crude Oil Prices: Weekly All Countries Spot Price FOB Weighted by Estimated Export Volume [Dollars per Barrel]. http://tonto.eia.doe.gov/dnav/pet/hist/ LeafHandler.ashx?n=PET&s=WTOTWORLD&f=W [accessed September 17, 2010].)

cubic foot from July 2007 through December 2009. For crude oil, the price nearly tripled from $50.1 per barrel in January 2007 to $137.1 per barrel in July 2008. It then also saw a huge drop, by a factor of nearly four, to $34.6 per barrel in January 2009.

The rationale about why natural gas and crude oil prices rose and fell as they did is certainly beyond the scope of this book and admittedly outside the expertise of this author. It is also a sure bet that not many people can explain or accurately predict when the price of these fossil fuels will rise or just how far they will tumble when they inevitably top out. Fossil fuels are traded on commodity exchanges, and their price is somewhat akin to the price of corporate stock on the stock exchanges that exist in a few major cities around the world. Just like the stock market, there is much analysis, discussion, and prediction about how fuel prices will behave in the future, but inevitably there are surprises.

We cannot respond quickly to these surprises nor can we really do much to mitigate their short-term impacts. It is not a simple thing to switch from one energy source to another. Your gasoline-powered auto cannot be easily converted to use wood as its fuel source. Even if the technology existed to perform this conversion, it would have to make sense economically. It might be simpler and safer to simply wait for crude oil prices to come back down. The entire price swing of crude oil described previously occurred within a two-year window. It certainly was financially painful for many to endure this steep increase in cost, but most did not have many options other than to try to be more frugal with their driving habits or perhaps to purchase a more efficient car if they were financially able to do so.

On a somewhat larger scale, an electricity-generating station burning millions of cubic feet of natural gas every hour cannot easily evolve into a sun-harvesting solar power center. The power plant was probably financially justified based on operating for several decades. Perhaps it was a peaking plant specifically designed to generate its power during the summer months when electricity demand tends to be higher, and coincidentally just when gas prices were hitting their maximum. Large utility electricity-generating stations were designed and constructed to meet certain environmental laws and regulations and may not be technically capable of operating in a significantly different fashion. They are a large fixed investment, capable of modification within a relatively small range of operating conditions, and not very nimble in the short term.

So, the amount of years left before our fossil fuels run out is not a very good basis for making decisions about energy use. As noted previously, it is somewhat inaccurate and does not necessarily reflect our local realities. What good is all the oil buried under an Arabian desert if I cannot get it into the gasoline tank of my car as it sits at a gas station in the middle of America? Supply, availability, price, and environmental impacts of fossil fuels are all important considerations.

Practical environmentalism is less concerned about how many years we have before the gas or oil runs out and, instead, concentrates on the impact our actions have to natural resources in general. Categorizing natural resources into perpetual, renewable, and nonrenewable is helpful in this regard as it allows us to better compare resource choices to identify which ones are most beneficial from a resource conservation perspective. Admittedly, this categorization of natural resources is somewhat academic, but it presents a good framework to begin to make choices and decisions about which type of resource we prefer to employ. In general, Practical environmentalism favors any action that tends to require less resources or make resources last longer, particularly with respect to nonrenewable resources.

There are two main ways that we can accomplish these objectives. The first is that we can increase the efficiency of the systems and processes that use natural resources. By employing smarter designs, we can simply use less resources to accomplish the same objective. We can design our vehicles to be more efficient in their use of fuel. We can design our factories to be more efficient in their use of electricity. We can design our homes to be more efficient with respect to the energy they require to keep us warm in the winter and cool in the summer. All these types of activities and choices put relatively less demand on the natural resources that sustain us. They also decrease the expense of purchasing the resource in the first place. Less fuel to run our cars, less power to operate our factories, and less energy to heat and cool our homes translates to less money spent on these necessities and more money available for other investments.

Resource conservation through improvements in efficiency tends to make economic sense. It is also quite predictable in that we usually know what initial costs, if any, are required to achieve the sought-after improvement in efficiency. We can compare the upgrade cost with the expected benefit and make a logical decision regarding the worthiness of the investment. It is a relatively low-risk process that allows us to plan for efficiency upgrades within the other constraints of our home or business budget.

I used this logic a number of years ago when I bought a new car. As you can probably tell by reading thus far, I am very interested in efficiency and resource conservation, and this interest spilled over into my decision for a new automobile. I chose a car that averages 45–50 miles per gallon (19–21 kilometers/liter) in fuel economy. You might expect that my car must be very small, even tiny. It is not. It seats four people comfortably, five if the backseat passengers are young or skinny, preferably both. I could have purchased a vehicle of similar size that would have given 20–25 miles per gallon (9–11 kilometers/liter) of fuel used. It would have accelerated somewhat faster and would have been slightly less expensive. I calculated the purchase price of both alternatives plus the fuel and maintenance expense to drive the car, and saw that the more efficient vehicle would cost less overall after only a few years. The increased efficiency, measured by lower fuel consumption, paid for the initial cost premium in relatively short order. Even now, after 10 years of operation, my car is still twice as efficient as many cars on the road and consumes fossil fuel at half the rate. The resource conservation pillar of practical environmentalism would record this as a win. Perhaps it is not the very best alternative. My car does still consume fossil fuel after all. However, my choice in automobiles still tends to preserve fossil resources more ably than other automotive alternatives.

Improvements in efficiency also tend to have a positive impact on the infrastructure that we depend upon to deliver natural resources to us. The water that spews forth from my kitchen faucet, for example, was drawn from an underground aquifer, treated to reduce the risk of biological contamination, stored in a large aboveground holding tank, and then delivered through a series of underground pipes to reach my home. All my neighbors receive their water in the same manner. Our town's water system has a certain fixed capacity to supply water to its residents. If we all use more water, or if our town grows as more people move into our lovely community, there may come a time when our water system will not be able to keep up with the demand. In this situation, our town would be forced to upgrade the water system. More pumps, more water towers, more treatment facilities, or more underground water mains might be required. This can be very expensive. However, if instead we use water more efficiently and reduce our consumption, then we postpone the expense of upgrading the system. Maybe we could postpone it indefinitely and enjoy very reasonably priced water for years to come.

This scenario of limited capacity to infrastructure and transport systems is present in many other areas as well. Sewage treatment systems, electricity distribution lines, natural gas pipelines, railroads, and highways all can experience limitations in capacity that require large investments to be able to satisfy increasing demands. Constructing new sewage treatment works, high-voltage power lines, or interstate highways is neither cheap nor easy. These improvements tend to be rather complex and involve public entities to direct, or at least permit, their construction. As such, they also usually take years to complete. So, not only does resource conservation have an immediate benefit of reducing the monthly bill for the resource itself but it also has a hidden benefit of postponing the cost of increasing system capacity that tends to keep the unit price of the resource from escalating at a rapid rate.

The second strategy to minimize usage of fossil resources is to substitute renewable or perpetual resources in their place. Using solar panels or windmills to generate electricity is a good example of this strategy. Both use renewable/perpetual resources to lessen the demand for fossil fuels to be burned to make electrical power. Every turn of the windmill blade and every minute the solar panel receives sunlight contributes to less coal, gas, or oil being consumed at an electric-generating station. Every hydroelectric dam that is constructed reduces the need to build a new fossil-fuel-fired power plant. Certainly the use of renewable and perpetual resources is not without impacts of their own, but from a resource conservation perspective, their use is generally favorable. Other impacts, both positive and negative, can be accounted for within the other three pillars of the practical environmentalism metric.

Earlier in this chapter we touched on the interrelationship between resource conservation and environmental degradation. We noted that wasteful overuse of natural resources can negatively impact the environment as well as our economic and social structures. We have now also described how the efficient use of natural resources can lead to beneficial economic impacts.

The resource conservation pillar is meant to help us translate these understandings into a form that has meaning on a personal level. Individually we can do very little to influence how many years of oil reserves exist or how best to manage radioactive waste material generated at nuclear power plants. However, we can make an informed judgment about whether our actions tend to conserve natural resources or waste them.

The resource conservation pillar is designed identically to the environmental degradation pillar and ranges from −2 through +2. Negative scores reflect inefficient or wasteful use of natural resources, while positive scores represent conservation or efficient uses of resources. As before, we have a choice between assigning an action with a value of one or two for the resource conservation pillar, which gives us some latitude in judging the size of the impact. Also as before, zero indicates no impact or a slight impact with a great uncertainty.

A negative score diminishes nature's capacity to provide resources to humanity, while a positive score improves this capacity. Actions or decisions that lead to increased consumption of fossil fuels deserves a negative score as this depletes the balance in one of nature's accounts, a very large account but one with a very low rate of interest. Projects that propose using renewable energy instead of burning fossil fuels deserves a positive score as they allow us to meet our energy needs without drawing down our reserves of nonrenewable resources. Programs to protect fisheries or watersheds also deserve a positive score as they protect the balance of these natural resource accounts and allow nature to continue to generate reasonable rates of interest in the form of continuous supplies of catchable edible fish and clean water.

Taken together with the scores from the first pillar, we now have a combined score representing the impact of our action upon the natural world. As you will see in a couple of chapters, you can simply add the two scores together using simple elementary addition. We will apply similar logic for the human pillars and eventually arrive at a holistic guide that helps us decide if our actions are likely to prove beneficial. In the next chapter we will discuss the human pillars of economic progress and personal benefit and incorporate them into the framework of practical environmentalism.

REFERENCES

Bagheera. 2010. The Dodo Bird Extinct in the Wild. http://www.bagheera.com/inthewild/ext_dodobird.htm (accessed September 12, 2010).

Energy Information Administration (EIA). 2010. http://www.eia.doe.gov/cneaf/electricity/epm/table1_1.html (accessed October 4, 2010).

International Energy Agency (IEA). 2010. Key World Energy Statistics. Paris.

World Nuclear Association. 2010. Some Physics of Uranium. http://www.world-nuclear.org/education/phys.htm (accessed October 4, 2010).

Energy Information Administration (EIA). 2010-2. U.S. Dry Natural Gas Proved Reserves. http://tonto.eia.doe.gov/dnav/ng/hist/rngr11nus_1a.htm (accessed September 15, 2010).

Energy Information Administration (EIA). 2010-3. World Crude Oil Prices: Weekly All Countries Spot Price FOB Weighted by Estimated Export Volume (Dollars per Barrel). http://tonto.eia.doe.gov/dnav/pet/hist/LeafHandler.ashx?n=PET&s=WTOTWORLD&f=W (accessed September 17, 2010).

7 The "Human" Pillars of Economic Progress and Personal Benefit

The third and fourth pillars of economic progress and personal benefit are meant to introduce a reality check into our environmental decision making. Practical environmentalism is greater and more inclusive than just the environmental issue itself. These two pillars are designed to force us to consider the realities of our livelihood and the strength of our preferences as we consider environmental action.

The human pillars represent the unstoppable forces of our human ambitions. These ambitions live within us and pervade our daily activities. We strive to "get ahead" both individually and collectively. Our governments set policies and enact regulations with the ultimate goal of strengthening national economies and increasing the value of the goods and services we produce. Individually we struggle to raise our standard of living, to become wealthy and achieve some measure of security. Many seek fame and fortune to be their life's greatest achievement, and we often define success in terms of wealth and popularity.

It is ultimately unworkable and unreasonable to pretend that human nature and human ambition have no place in environmental decision making. The vast majority of us do not live and act in some altruistic vacuum. The few who do end up with the title "Saint" in front of their names.

The goal of including these two pillars in the metric is to encourage environmental action that has a reasonable chance of success. Why should we waste precious time and effort advocating some position, program, or activity that will ultimately fail because a majority of our society will not support it? Even if we plan some small action that affects or involves no one but ourselves, why should we pursue it if it will drain our wealth and has negative personal benefit? Admittedly, there will be cases where some of the pillars are positive and the others negative. We may be faced at times with choices of compromise or sacrifice. We will discuss how the pillars deal with these situations in later chapters but, for now, we will focus on representing the "human" pillars within practical environmentalism.

ECONOMIC PROGRESS

Economic progress has been the historical adversary of environmental protection. Throughout most of humanity's history, economic progress has been tied to our ability to provide food for ourselves. A rather nomadic existence of hunting and gathering evolved to a more stationary agrarian lifestyle. We began to view particular plots of ground as "our own," and with each spadeful of earth turned over in our labor to survive, we changed the earth. Virgin prairies protected by thick sod were laid open and exposed to wind and water. Ancient forests were laid low to provide fuel to warm us, timber to shelter us, and even more space for us to plant seeds and grow more food.

In most cases, nature tolerated this intrusion rather well. Nature is versatile, flexible, and very patient. Nature has a vast number of life-forms in its inventory and will push forth whichever local species has the best opportunity to thrive. In many cases, nature bent to our will, and the plants and animals humanity desired generally flourished. Occasionally, Mother Earth would wipe away our best efforts by sending flood, fire, drought, or pest, perhaps as an explicit reminder of the true balance of power.

As we humans learned how to take advantage of nature, we began to profit from nature. We began to produce more than we actually required to sustain ourselves. We produced excess grain and bred herds of livestock. We planted fruits and vegetables, and harvested more than we could consume. We learned that nature's desire for growth was very great indeed. Life itself was yearning to expand. Humanity learned how to remove a few of nature's checks and balances in order to allow certain desirable species to outcompete and outperform less desirable ones. Grains replaced grass and weeds. Cattle and sheep and swine were guarded and protected from nature's predators and allowed to breed and multiply. Nature's tendency to grant life yielded wealth for us.

We began to trade our excess production for other goods that we desired. Little by little our clothing became finer, and our homes became grander. We learned how to make metal and shape it to our needs. We manufactured better tools that made it easier to plow our land and harvest our crops. We became wealthier.

As our riches grew so did our time. No longer did we toil all through the day in hopes of having enough food to live another day, week, or month. Our labors were no longer forced to end when darkness came, nor were they limited by the extent of our physical strength and energy. Instead, we gained the luxury of ceasing our labors according to our desires. Our ambitions could be married to our efforts. "Free" time was created, and we could choose how we spent it. We could continue our work and strive to be rich. We could entertain ourselves through stories, music, or art. We could seek after knowledge and become learned. We could travel beyond our immediate surroundings and explore the world.

We began to employ ourselves in occupations other than growing food, and we began to congregate in villages, towns, and cities. Our labors became specialized and more efficient, and we distanced ourselves from the land that heretofore had directly sustained us. Trades and professions appeared, and gradually it became less acceptable to soil your hands in Mother Earth's dirt. Aspiring to be a physician or lawyer or professor was preferred to being a farmer. Our societies created hierarchies of labor and occupation, and those associated with the land migrated to lower levels of status. We started to separate ourselves from nature, and the connection between our prosperity and nature's generosity began to dim in humanity's consciousness.

The industrial revolution that began toward the end of the eighteenth century was a paradigm shift in the evolution of economic progress and accelerated the pace of our emigration from the land. Humanity finally figured out a way to combine knowledge, tools, and nature's energy to create machines capable of laboring at profound rates. Once again, nature's generosity was the key to this transformation as it multiplied human effort manyfold. Abundant energy existed in the rushing waters, gusting winds, and the heat liberated from wood and coal. Water wheels powered the mills that sprung up along the riverbanks. Windmills pumped water from underground aquifers to quench the thirst of ranchers' livestock on arid plains. Wood and coal were no longer confined to the hearth but became the driving force for steam engines and boilers that powered industry and transport.

The industrialization that the revolution brought changed the way humans worked. Many no longer struggled with the plow cutting through the land or worried if there would be enough rain at the right time to prevent their crops from withering in the field. Instead, we worked with machines and struggled with them. We took raw materials, applied our labor in concert with the machines and manufactured products. More and more people moved indoors to satisfy the demands of industry. Industry needed us to man the machines. We were paid for our labor and no longer subject to the whims of nature. Now we were subject to our employer. Figure 7.1 symbolizes humanity's relocation of work and effort from the great outdoors to the new confines of industry.

FIGURE 7.1 Industrialization. (Illustrated by Mark Benesh. With permission.)

I have seen little evidence to suggest that we paid nature its rightful due in making the industrial revolution possible. While humanity made it practical, nature was still the driving force powering the machines. We humans are limited in energy, while nature provides it in abundance. A human being is weak compared to an ox or horse when it comes to plowing a field. We are confident that our intellect is superior, but our muscles are flimsy and inferior by comparison. We are slower than a deer and bound to the land unlike the hawk or eagle. We rarely produce more than one off-spring at a time, and then must tend to our children constantly for years to enable their survival. In fact, many of Mother Nature's species are superior to us in one way or another. Our advantage, of course, is our individual and collective mind, and our capacity to reason, deduce, and learn. Brainpower has proved itself superior to horsepower.

Yet for all our brainpower, we generally failed to make the connection that our economic proficiency and prowess was fundamentally the result of nature's bounty. Nor did we realize that we, especially as an industrialized society, could have an impact on nature that could be detrimental to both nature and to ourselves. As a society, we were caught napping as to how much power industrialized processes had to impact their surroundings. Our machines could belch pollutant-laden smoke into the sky at very rapid rates. Our factories could leach filthy odorous water back into the rivers and streams at rates great enough to overwhelm nature's capacity to cleanse them before we again drew from them to drink. We enjoyed newfound productivity and wealth but were too ignorant or paralyzed to recognize the detriments to our natural environment. We had unleashed a monster and did not even know it.

This monster lives with us today, hopefully tamed and restrained, but certainly present with us. Industrialization is not likely to go away anytime soon. Factories, office buildings, and their supporting infrastructure are here to stay. They are part of our modern way of life even as we strive to keep check on industrialization's capacity to degrade the environment.

This brief semihistorical rendering of humanity's economic evolution is meant to highlight how ingrained prosperity is within our being. Over centuries and millennia, we humans have evolved an inner drive for security, status, and wealth. It is truly part of us. Religion may argue that this drive for prosperity is an undesirable part of our being, that it prevents us from becoming fully enlightened, or that it hinders our salvation. I am certainly not a theologian, nor do I wish to enter into a debate on such matters here. When it comes to the propriety of our desire for wealth and power with relation to the state of our souls, I will defer to others to make the arguments for and against. I do believe and will argue, however, that this desire to improve ourselves economically, whether ultimately good or bad from a religious perspective, is present within us and ultimately good for our present well-being. Economic progress certainly has a dark side, but a light one, too. Prosperity has allowed us the time and opportunity to pursue the finer things in life. It has graced us with health and a longer life span. It has allowed us to provide for our children and given us the joy of watching them prosper themselves. Our lives are more than a continuous struggle for sustenance and gratefully so. Whether we wish for leisure or challenge, for great works of art or fine, fast automobiles, for a comfortable home or travels beyond the horizon, we are fortunate to at least have the opportunity.

Economic progress granted us these wishes, and for most of our history did so at negligible cost to the environment. It was not until relatively recently that economic progress and the environment from which it came started to appear in stark contrast to one another. Even so, this conflict is not reason enough to abandon economic progress as a desirable, worthwhile pursuit, again with my apologies to the spiritualists. From a purely practical point of view, economic progress is a good thing and worthy of inclusion as a pillar of our environmental decision-making framework.

It is very interesting to consider that over the decades of the twentieth century, slowly but surely, our industrial processes have become more efficient and less detrimental to the environment per unit of production. Economies of scale have prodded us to build bigger and bigger factories that produce goods in larger quantities. In general, economies of scale allow us to use less material, less energy, and often fewer workers, and generate less waste with every individual component manufactured. These efficiencies tend to lower manufacturing costs and allow new large facilities to outperform and outcompete older smaller factories. Along with new factories has come the general requirement from governmental regulatory bodies to install and employ pollution control equipment. Of late, recycled materials are used more and more in many products and reduce the toll of extracting virgin materials from Mother Earth. Hazardous materials have come to be regulated and monitored, and companies that use them have strong incentives to minimize their usage and even change their manufacturing and product designs to eliminate their use altogether. Scores of environmental advocacy groups have sprouted up and exist today, keeping a watchful eye on industry and government to make sure that economic progress does not result in undue stress to the environment.

And all this occurred while economic progress continued. Some might make the claim that economic progress allowed these advancements to occur and that wealth paid the way for environmental betterment. Either way, we are starting to realize that economic progress and environmental health are not necessarily contradictory. While some industry might bear the cost of increased environmental protection, other industries profit in the new economic and environmental landscape. Preventing excess pollution from entering a water body, for example, could benefit local tourism and fishing industries. It also benefits the engineering firms and construction companies that design and install the new pollution control equipment. State and federal mandates for renewable electricity generation creates incentives for corporations to build more and better wind turbines and in my home state creates income streams for farmers who lease their land for the operation of these giant machines that turn the movement of the atmosphere into electricity that powers my toaster and my television.

The question, of course, is balance. If some action to safeguard the environment requires an economic penalty to some industry, we must decide whether that penalty is offset by benefit in other areas. This is not an easy task in many instances. Teams of economists and engineers may be required to work through a complex question. A method known as "cost-benefit analysis" is often employed to quantify and compare the costs and benefits of some action in order to provide guidance to decision makers. It can be difficult at times to quantify or monetize these costs and benefits, and some particular questions are steeped in value judgments. How much is a snail darter worth, for example? How valuable is air that is cleaner and water that is more pure? Should we really use acres of prime farmland as a dump to bury our trash?

While the details may be tricky, and a perfectly accurate cost-benefit analysis may be impossible, there is considerable value in the imperfect execution of the method. If we allow ourselves to do it "well enough," it does not have to be that hard. In fact, most of us perform cost-benefit analyses daily without giving it much thought. Just about every purchasing decision we make includes an implicit cost-benefit analysis. Do we order prime rib at the restaurant or a hamburger? Does our cable TV service have every movie channel under the sun, or is it merely "basic cable?" (I am a basic cable proponent myself, and my kids do not appreciate it much.) Do we buy the latest computer technology or wait till it has been around a while? All of these decisions, and many more, have a cost-benefit component to them, and the benefit may not necessarily be easy to quantify. It is, however, part of our daily economic experience. It is part of our human nature.

So if we can perform a basic cost-benefit analysis as effortlessly as ordering dinner at our favorite restaurant, we can choose an appropriate score for the economic progress pillar. The score range is just like before, running from −2 to +2 with zero sandwiched in the middle. Negative scores represent actions that tend to damage our economic systems. Positive scores represent overall economic benefit, and zero is a toss-up or no impact.

Now I realize that this might seem like a daunting task, and this pillar is probably the toughest for a lot of us to evaluate with a high degree of confidence. Most of us are not economists, and even the smartest economists do not have the right answer all the time. Consider the volatility in the stock market and the financial crises that seem to befall our economies from time to time. Who can explain all that? Who can predict these economic disturbances and guide us to prosperity? The point here is that when the economic situation is complex, there are no easy answers. To use the matrix, however, we do not need a perfect answer, or even one that is highly accurate. All we need is a sense whether the proposed action will likely result in a negative or positive impact to our economy. Once we decide positive or negative, then we choose between a large impact or a small one. If it is a simple environmental issue, affecting only ourselves or our immediate surroundings, then our economy is literally our personal wealth. If it is a large environmental issue coupled with a proposed large response, similar to a new federal law, for example, then our economy is our national economy.

However, it is very fair to score this pillar as it relates to each of us personally regardless of the size of the issue. Big issues are big because they impact lots of individuals. Each of us could be personally impacted by some big environmental law. We are certainly entitled to form a judgment based on personal economic impact. This is an easier application of the economic progress pillar and one that is more aligned with the intent of this book. This is not a book about how experts should evaluate governmental legislation. This is a book about how you and I can make decisions about how we should individually respond to the environmental issues in our lives. I will bet that we know enough, or can learn enough, to make a reasonable judgment for the economic progress pillar.

PERSONAL BENEFIT

The fourth pillar, more than any of the others, is highly focused on the individual using the metric. It is entitled "personal benefit" after all. It is included to allow us to consider human-centered impacts that are not strictly economic. It allows us to incorporate our own values, ethics, and personal judgments explicitly into our environmental decision making. We would likely do this anyway; we might as well do it rationally and with forethought.

While economics is crucial in our daily lives, few of us live solely to generate the maximum amount of wealth we can just to park it in our bank account. Instead, for most of us, wealth allows us to live our dreams. We want to travel, or perform on stage, or learn a new language, or be in great shape, or grow a beautiful garden, or watch our kids grow up. We want to be healthy, to live a long and vibrant life, and to have fun. Money and wealth is often merely a way to get what we really want.

We human beings have very strong personal preferences at times. We grow from childhood into adulthood in a process that gradually defines what we like and what we do not. Our preferences shape our personalities and often dictate where we live, who we associate with, and how we enjoy our free time. We are unique creatures and can easily spot the differences and diversity among us. Even twin children, seemingly identical in appearance, are obviously very separate and distinct people when you spend a little time with them. Two good friends from my youth were twins. We kids never had any trouble distinguishing the two. While they looked very much alike, they were not perfectly identical. At first glance we knew who it was. We also knew which games they each individually liked and which they would really rather not play. We knew who was better at sports and who preferred them least. They were certainly individuals to us.

Our preferences are strongly shaped by our histories. Our nationalities, cultures, languages, and geographies all play a part in influencing what we value. Sometimes this influence is subtle, and our preferences become attached to us without our noticing. We spend our childhood in a particular cultural environment, and it stays with us forever. Our religious faith is typically the faith of our parents. Our attitudes are often founded on their attitudes. Our diversions can be traced back to our childhood and what our parents enjoyed, or allowed us to enjoy.

I grew up in a suburb of Chicago, Illinois. During my youth, my father was a professor of education and taught in downtown Chicago. My mother was a homemaker and cared for my brother, sister, and me. My father enjoyed games and sports, and we spent many fun hours in those activities. I enjoy games and sports to this day. My mother enjoyed gardening and living in a neat well-organized home, things that I now value too.

There are preferences and traits within me that I cannot trace back to my parents, however. For one, I love automobiles, while my father is quite happy not needing one. He is a fan of mass transit and can get just about anywhere he needs to go without climbing behind the wheel of his personal automobile. My dad is an avid walker and also knows how to utilize the bus, train, and airplane to his advantage. Now, I have nothing against public transit, in fact, I think it is very desirable and beneficial. I too am a big fan of walking and appreciate living in a quaint Iowa town where I can walk to the grocery store, post office, hardware store, café, and movie theater. I made good use of the buses in Iowa City during my time in graduate school there, and I certainly prefer to take the train to get to Wrigley Field to see my beloved Cubs. But I really like cars too, always have from what my parents tell me. My mother says that this particular trait must have come from my grandfather who was enthralled with the automobile as well. While my love of cars was not shared by my parents, they certainly allowed me some latitude in pursuing my interest. They put up with me owning more cars than they while I was still in high school and allowed me to spend a good amount of time with a family friend, a mechanic by trade, who taught me much. And now, in my adulthood, I still have a deep interest and affection for this particular form of four-wheeled personal transportation. And, lucky for me, my youngest son really likes cars too, so I will have a good excuse to continue having automobiles in my life for a long time. Maybe we will tinker with one somewhat like that shown in Figure 7.2. I fully expect automobiles to continue to change and evolve, and hopefully in an environmentally conscious and friendly manner. So whether you attribute it to nature or nurture or both, my preference in favor of the automobile is strong, and I will naturally incorporate it within my own environmental decision making.

Living in Iowa, I have run across many who enjoy hunting and fishing. While I personally have nothing against these activities, they are not among my preferred ways of spending my time. Perhaps things would have been different had I grown up in a more rural environment with a father or uncle or brother or sister who cherished these activities, but perhaps not. Today, it is not very difficult to find people who are passionate on both sides of these issues, hunting probably more so than fishing. I imagine that the traditions of our upbringing, our family, our place, and our culture can have much to do with these preferences.

Certainly as teenagers and young adults, we may purposefully chart our own, sometimes contentious, courses to our lives. But, our youthful experiences of family, friends, school, and hometown are indelibly imprinted upon us in some form or fashion. Whether we embrace our history or flee from it, its memory is within us and influences our preferences today.

Sometimes we have a life-changing experience, and we are suddenly and abruptly aware of an issue that we care deeply about. We may make a new friend that introduces us to a new way of thinking about a particular issue. We may be confronted by disease

FIGURE 7.2 My son's first car? (Illustrated by Mark Benesh. With permission.)

or hardship and develop a new appreciation for our relationships with others or even with nature. We may witness some event so powerful that we are forever changed.

Our experiences, whether gradual or sudden, influence our preferences and can influence the way we think about the environment. Our preferences are generally not conceived in our minds but rather born in our hearts. The personal benefit pillar is designed to capture the strength and passion of our emotions. It recognizes that what we feel individually is important and deserves explicit recognition. It acknowledges that environmental issues are often tightly wrapped in emotion and grants us a way to give voice to the emotion. Sometimes we feel things very strongly before we can make sense of them mentally.

The fourth pillar recognizes that our feelings and desires may be intertwined with the environmental issue at hand or the action we consider. We may be quite willing to accept a score of −1 on the economic progress pillar if it comes with a score of +2 on the personal benefit pillar. An avid fisherman recognizes how important water quality is to what he loves to do and may be very willing to pay taxes to support construction, operation, and maintenance of the sewage treatment plants that keep our rivers and lakes clean and biologically healthy. A hiker and nature lover probably values large national parks and is willing to forgo the economic return from developing the resources on those lands.

Similar to the other pillars, the personal benefit pillar is constrained to integer values between −2 and +2. Once again, we have the flexibility to rate our proposed action either positively, negatively, or neutrally, and with weak or strong responses on both positive and negative scales. Limiting the score of the personal benefit to this range may feel somewhat restrictive, especially for this pillar. We often feel things so strongly that we may become single-mindedly driven to accomplish our objectives. When I consider the many bold actions undertaken in the name of environmental protection, I tend to think that those actions may have more to do with personal values and ethics than a cold-reasoned analysis of environmental degradation. Those who invest so much time and energy into their cause must be deeply invested in it on an emotional level as well. I am not implying that their views are necessarily extreme or unreasonable. Rather I am suggesting that their own personal values are tightly linked to the environmental issue that stirs their passion.

As mentioned in the introductory chapter to this book, the four pillars are specifically designed so that no pillar overwhelms another. Practical environmentalism is about a balanced approach to our analysis and response to environmental issues. It is not meant to be a coldly analytical methodology devoid of our humanity. It does specifically include the personal benefit pillar to account for our feelings and emotions, and allow them their proper place in our decisions. It does not, however, allow our feelings to rule our judgment. Practical environmentalism calls for emotion to be tempered with reason, and for logic to be seasoned with feeling. Practical environmentalism, through the structure of the pillars themselves, as well as the scoring of each, leads us toward rational comparative analyses of the four factors that have so much to do with how we respond to perceived environmental calamities.

The metric is made up of four pillars, each to be used and judged individually by you and me. Neither emotion, economics, resource preservation nor environmental damage is allowed to rule over the other three. They are all important and taken together can hopefully produce decisions and results that provide environmental benefit that we can accept.

Now you have made it to the midway point of this book. Hopefully, the groundwork of practical environmentalism has been clearly laid. We have discussed the history and ethics of environmentalism, as well as some of the confusion that swirls around our attempts at environmental betterment. We have examined each pillar in detail and discussed how and why each is included within the practical environmentalism metric. Now it is time to begin to put all this work and study to good use. In the next chapter, we will examine how to combine the four pillars both quantitatively and qualitatively to guide our environmental decision making. Then in Chapters 9 through 12, we will put the pillars to the test with examples from the past and present, and see just how practical environmentalism casts a new light on our environmental decision making.

Let us start putting the pillars back together.

8 Scoring with the Pillars—A Few Simple Examples to Illustrate the Method

The power of practical environmentalism lies in its ability to deconstruct an environmental action into its component parts, analyze them rationally and reasonably, and then put them back together to form a holistic judgment of environmental benefit. The previous three chapters have performed the task of deconstruction. We have examined each pillar in detail, hopefully reached a better understanding of each, and learned how to use them when considering environmental action. Remember that practical environmentalism is action focused. Remember, also, that it presumes that we care about the environment and want to take some level of action to protect it, even improve it. So the four pillars that make up the metric are meant to be applied to actions that we consider, or choices that we make, with the ultimate goal of environmental betterment. Our goal in this chapter is to begin to understand the process of reconstructing the four pillars into a single judgment that can guide our choices and actions.

There are a couple of reasonable ways that we can combine the pillars within the metric. First, we can simply look at whether we assigned a positive, negative, or zero value to each of the pillars and let majority rule. This is the qualitative approach, and it is very simple and basic. If combining the four pillar scores yields more positives than negatives, then our judgment should be to proceed with caution. If the combination yields more negatives than positives, then we should put on the brakes and rethink our action. The following two tables show several generic examples of how we might combine scores to indicate whether to proceed (Tables 8.1 and 8.2).

Note that while this table shows all positive ratings for the environmental degradation pillar, this need not necessarily be the case. It was shown this way simply to reflect an expectation that environmentalists typically do not look for ways to degrade the environment. We usually consider actions that will improve the environment or, at the very least, reduce the rate or scope of potential degradation. It is conceivable, however, to use the metric to evaluate situations where the environmental degradation pillar is negative, but some or all the other pillars are positive, and the metric results in a positive indication. This is perhaps the "business as usual" case where we accept some measure of environmental degradation as a trade-off for economic gain or personal well-being. Here is another table showing how individual pillar results can be combined, in this case, to produce negative or neutral total scores with the majority rules approach (Table 8.2).

TABLE 8.1

Examples of Positive Combined Scores

Pillar	Example 1	Example 2	Example 3	Example 4
Environmental degradation	+	+	+	+
Resource conservation	+	−	−	0
Economic progress	+	+	0	0
Personal benefit	+	+	+	0
Combined score	+	+	+	+
Indication		Proceed with caution		

TABLE 8.2

Examples of Negative and Neutral Combined Scores

Pillar	Example 1	Example 2	Example 3	Example 4
Environmental degradation	+	+	+	+
Resource conservation	−	−	−	−
Economic progress	−	−	0	−
Personal benefit	0	+	0	−
Combined score	−	0	0	−
Indication		Whoa! Let us think about this some more.		

Again, the scores for the environmental degradation pillar are shown as all positives for the same reason as stated previously. Now though these environmental benefits are offset by penalties in the other pillars. In none of the examples shown in this second table do positive scores outweigh negative ones. Here, we must reign in our desires to save the world because we may be doing more harm than good; more to come about that later.

While I am arguing for the use of the metric as a way to increase our understanding and consideration of environmental issues, I cannot preclude its use to only those issues that have demonstrated environmental benefit. In fact, I would hope that practical environmentalism metric would be broadly applied in "business as usual" situations and in so doing would encourage more thought and consideration be applied to the environmental impacts of our actions in general. Practical environmentalism does not seek to limit its application to only those efforts that are done expressly to protect the environment. It seeks rather to incorporate environmentalism within our daily affairs. It certainly acknowledges the importance of economic considerations and personal values within the choices we make, while at the same time calling for environmental degradation and resource conservation to have a place within our decision making.

The second way to combine the pillars is to simply substitute numeric scores for the positive, negative, or zero scores used earlier. This allows us to be more quantitative in our analysis and allows us to take advantage of the differences between

TABLE 8.3
Examples of Quantitative Analysis

Pillar	Example 1	Example 2	Example 3	Example 4
Environmental degradation	+2	+1	+2	+1
Resource conservation	−1	−1	+2	−2
Economic progress	0	−2	−1	0
Personal benefit	+1	+1	+2	+1
Total score	+2	−1	+5	0
Indication	Proceed with caution	Whoa!	Proceed with caution	Whoa!

large and small impacts. I like the sense of security and accuracy that comes with the knowledge that one plus two equals three. Table 8.3 shows a few examples of this method.

We now include a row totalizing the scores from each of the four pillars. Decision making remains essentially the same with positive totals indicating that it is OK to proceed, and negative or zero totals indicating that we should rethink things. This quantitative method also gives us the ability to judge our actions based on the magnitude of the total score. Greater absolute values, both positive and negative, lead us toward easier decisions. Example 3 in the preceding table reflects this situation with a total score of +5. That is a very clear indication that we believe our action will result in a beneficial result overall. Total scores of +1 or +2 are relatively marginal and certainly deserve more caution as we move forward.

Both methods illustrated are viable and reasonable ways to go about environmental decision making. You can choose whichever method is the most comfortable to use. It should be noted, however, that the qualitative, positive/negative, majority rule approach tends to be more conservative than the quantitative analysis method. That is to say that majority rule approach can produce neutral results when quantitative analysis indicates a positive result. The qualitative approach ignores the weighting of both positive and negative pillar scores. A +1 score is no different than a +2 score in a qualitative analysis, and neither is there any difference between a −1 and −2 in this approach. Let us look at an example to illustrate this point.

We can imagine that we have discovered a radioactive chemical leaking from metal drums buried underground at an abandoned factory site. The factory owner is long gone, of course, and the factory site sits as an ugly eyesore to the community. Our discovery, however, has just increased the stakes. We theorize that the radioactive waste has leached into an underground aquifer and now threatens the local water supply. People are upset and frightened, and rightly so.

Being the good environmentalists we are, we naturally want to clean up this mess, safeguard the environment, and protect the local residents. It will not be easy though. The leaking chemical is very nasty stuff and not amenable to the more efficient technologies used to clean up underground spills. If we want to clean up this mess, we will have to dig up the site to remove the metal drums, and we will have to drill perimeter wells to try to intercept any radioactive material before it escapes into the aquifer. We will have to closely monitor the site and our cleanup crews for radiation exposure. It will be very expensive. Nevertheless, we are highly motivated to set things right, and we would like to proceed with the cleanup efforts. Such efforts would require a lot of money and a lot of involvement with the local community and appropriate levels of government. Public interest would likely be significant, and emotions could run very high. We had better do our homework in researching the problem and coming up with an effective solution. Let us use the practical environmentalism framework to assist us.

The environmental degradation pillar is pretty clear in this case. We decide that our proposed cleanup would prevent or mitigate significant environmental damage. We rate our plan as +2 in this regard. The resource conservation pillar is not so easy. The cleanup will take a lot of time and a lot of energy. We will have to use large mechanized equipment to dig up the site, and much of the contaminated earth will have to be treated to minimize its volume before transfer and disposal. The perimeter well field pumps will run 24 hours per day and 7 days per week. The water they pull from the ground will need to be treated as well. Our plan will consume significant amounts of electricity, diesel, and gasoline. Acknowledging this, we rate the resource conservation pillar as a −1.

Economic progress also gets a −1 rating. To be sure, some contractors are going to make a lot of money, and the electric utility and the fuel supplier will appreciate the business. However, we are not producing goods or services that are generally of value to society. We are cleaning up a mess, a negative externality of a previous manufacturing operation, and it hinders our economy more than it helps. We must reserve positive scores for the economic progress pillar to situations where we produce goods and services that society wants and/or where we improve the efficiency or effectiveness of our economic enterprises. Waste cleanup is an unfortunate and negative, if sometimes necessary, economic endeavor overall.

The personal benefit pillar in this case depends heavily on your perspective, and this is, of course, quite typical. Naturally, people living close to the affected area would likely appreciate the cleanup effort, and a +2 rating would be reasonable. People living farther away would likely not rate it so positively, perhaps with a +1 or maybe even a 0 if they did not know about the problem or did not really care. You could argue, however, that there is some personal benefit, although small, to almost everyone in knowing that toxic waste will be removed from the water supply, and the danger to humanity will be reduced. Given that our cleanup would likely rely on public funds, it is appropriate to consider the entire range of personal benefit, and we could easily conclude that an average value, say +1, is appropriate. Table 8.4 shows these scores under both evaluation methods.

As you can see, the two methods produce two different results. Quantitative analysis results in a positive score and the recommendation to go ahead with the cleanup. The qualitative method produces a neutral result and the recommendation to rethink the situation. In fact, the qualitative approach will produce a neutral result any time that two pillars are positive, and the other two are negative. Considering the complexities of environmental issues today, this situation could happen quite frequently.

The zero score from the qualitative approach is a more conservative recommendation in that doing nothing is generally more conservative than doing something.

TABLE 8.4
Toxic Waste Cleanup Example

Pillar	Quantitative Analysis	Majority Rules
Environmental degradation	+2	+
Resource conservation	−1	−
Economic progress	−1	−
Personal benefit	+1	+
Total score	+1	0
Indication	Proceed with caution	Whoa!

Taking action of any kind is less conservative than doing nothing at all. For once we take action, we commit ourselves to change in some form or fashion, and change sometimes brings with it unexpected results or impacts. It is often quite difficult, if not impossible, to perfectly reverse our action in case something goes wrong.

You might ask what could go wrong with our little cleanup? Unfortunately, there are many things that could potentially go wrong. Hopefully, most of these potential negatives have a very low probability of occurrence, but they are possible. We could have an accident during our work and injure or kill one of our cleanup crew members. We could spill more of the radioactive chemical during the cleanup, either at the site or during transportation to its permanent and, hopefully, secure, storage location. We might do all this work and still have unacceptably high levels of radioactivity in the groundwater. It is for all these reasons that being conservative might mean undertaking a minor cleanup instead of a major one. It could also mean doing absolutely nothing.

You might also ask why qualitative analysis always tends to be more conservative. It is certainly true that it devalues +2 scores, but it would also devalue −2 scores, right? The Table 8.5 shows this situation.

Once again, the positive and negative scores are balanced two and two, and the majority rules approach delivers a zero score. In this case, for the quantitative approach, the −2 scores for resource conservation and economic progress outweigh and overwhelm the +1 scores for environmental degradation and personal benefit. This example shows a case where a relatively minor environmental improvement,

TABLE 8.5
Generic Example with Greater Negative Impact

Pillar	Quantitative Analysis	Majority Rules
Environmental degradation	+1	+
Resource conservation	−2	−
Economic progress	−2	−
Personal benefit	+1	+
Total score	−2	0
Indication	Whoa!	Whoa!

even one that delivers some level of personal benefit, is very costly in economic terms and in the use of natural resources.

The key to understanding the built-in conservatism is that both negative and neutral scores produce the same overall judgment. In this case, both approaches result in an indication not to proceed with the contemplated action. Certainly, there is a numerical difference between a −2 score for the quantitative approach and a zero for the qualitative approach, but they both ultimately lead us to the same conservative conclusion.

I am not necessarily advocating being conservative as a general rule. I am simply trying to make a small point about how the two methods can differ in certain situations. Ultimately, the score for each pillar and the method we use to reach a recommendation are up to those who contemplate action.

The metric is meant to help those who are proposing action, as well as to help those who evaluate, judge, and take positions on proposed actions. Used well, the metric will clarify environmental issues and clarify our own personal values with respect to these issues. And always remember that all positive indications that prompt us to action also come with the recommendation to proceed with caution. We must humbly recognize that we do not know all things, and we are not perfect predictors of the future. Realistically, we are not even fair predictors of the future! So, in all our environmental actions and responses, we should move ahead thoughtfully and carefully. And, bigger actions require more thought and more caution.

In the remainder of this book, we will neglect the qualitative approach in favor of the quantitative. I feel that inclusion of numerical values provides a superior analysis in that it allows us to better differentiate impacts both within and across the four pillars. It is also less conservative from an environmental perspective and tends to favor action over inaction. The environmentalist within me favors the call to action to improve our world.

Within the quantitative approach, it is fair to question why the numerical range is limited −2 through +2. If one of the strengths of the quantitative approach is to differentiate between greater and lesser impacts, then would not a range of −3 to +3 be even better? Or perhaps, −5 to +5 would be great, or −10 to +10 would be superior. It is easy to see that we could choose an unlimited number of numerical ranges to use as a basis for scoring each pillar. The greater the numerical range, however, introduces more difficulty in assigning scores as there are more choices to make, and the real difference between adjacent values becomes less and less. Within our current range of −2 to +2, there is a significant difference between a score of +1 and a score of +2, and it is relatively easy to differentiate between them. If we expanded our scoring range to −10 to +10, there would be a much smaller difference between a score of +1 and +2, and we would also have to consider whether the pillar score should really be +3, or even +4. How would we decide what the appropriate score should be when we have 10 positive scores from which to choose? In this scenario, would there really be much of a difference if we chose a score of +2 instead of +3? We would have to develop some logic, at least within our own minds, to differentiate between successive ratings. The more ratings we have, the more difficult it becomes to perform this differentiation.

Practical environmentalism is not about creating complexity for complexity's sake. It does seek an appropriate amount of consideration for the multiple factors represented by the four pillars, but also attempts to structure our analysis so that we can reach viable conclusions without undue effort. Limiting our scoring range from −2 to +2 is a workable compromise between the ability to differentiate impacts within the pillars and ease of use of the overall method. It really gives us a threefold range of scoring rankings with zero being little to no impact, 1 being a moderate impact, and 2 representing a large impact.

Having hopefully made the case for the −2 to +2 scoring range, you will find in later chapters that we do split hairs at times with scores near zero. In some examples, you will see a modified 0/−1 or 0/+1 score used. You can think of it as a half score if you wish, −1/2 or +1/2. This slight modification allows us to pick a side of zero for our score. This helps in the case where we believe the impact will be very slight, but decidedly to one side of zero or the other. For example, perhaps we are contemplating enacting a new environmental law that would force sewage treatment plants to upgrade their systems by installing additional equipment to reduce the levels of nitrogen in their effluent. For those of us not experienced in designing sewage treatment plants, it will be hard to know how much these improvements would cost and how much they might add to our monthly bill. We could invest a lot of time researching the issue and learning all about nitrogen removal from wastewater, or we could make an educated guess. It seems reasonable to assume that the additional equipment required will cost money. Not many companies are in business to install pollution control equipment for free. And, when was the last time you saw your utility rates decrease? It does not happen very often.

Maybe the sewage treatment plant in your town might qualify for government grants or special funding of some sort. That might help some but, at the end of the day, somebody will have to pay to upgrade the system. So it seems reasonable to believe that the economic progress pillar will be negative, just how negative though is questionable. Perhaps with just a little bit of research, we learn that the new equipment is not expected to be terribly expensive and that the cost increase to the average homeowner will be less than 1%. This does not seem like very much. We might not even notice the change on our bill. So, in this case, we can split hairs and say that while the economic impact will be negative it will be small, too small to justify a score of −1. We can apply the compromise score of 0/−1 and go on to rate the other pillars.

With many environmental issues we face, this ranking system will be more than sufficient. It gives us the capability to easily separate positive impacts from negative ones, greater impacts from lesser ones, and even to dance around the zero score. Remember that practical environmentalism seeks a direction rather than the answer. There are very few true "answers" to environmental dilemmas. What is the "solution" to global climate change, or acid rain or species extinction? What is the "answer" for nuclear waste or urban sprawl? These are trick questions, of course, questions in which a single answer does not really exist. There are many possible responses, some of which might ultimately prove beneficial, while others might generate unintended and unwanted consequences.

What we have instead are choices that tend to make the environment better off or worse off. These choices may also make our own lives better off or worse off. Practical environmentalism seeks to help us make these choices. It uses a simple structure that is easy to implement, yet thoughtful enough to prevent trivializing the process. The true force within the method is not the method itself, but rather the structure that allows the method to be used easily, well, and hopefully often.

In the chapters that follow, we will use this method over and over again with various examples. These examples will run the gamut from our own very personal choices about how we interact with and impact the environment on a daily basis, up to the global issues that we hear so much about and that seem way beyond our individual capacity to impact in a positive way. Our goal will be to become familiar and comfortable with the practical environmentalism method and learn how we can use it to influence our action, as well as inform our opinion. Let us continue by examining how we can use the pillars in our own daily activities.

9 The Pillars in Daily Life

Environmental decisions present themselves every day in forms large and small. Should I walk to work or school today? What about global warming, should I write my congressman and demand that the government do something? Should I buy some solar panels and put them on my roof? Should I get a different car or turn up the temperature on the air conditioner or get water-saving fixtures for my bathroom? The pillars can be used to address these and most other environmental choices that cross our minds.

This chapter is really the essence of the book. Here is where we can make a difference all by ourselves. We do not need government mandates or direction from the pro-nature environmental advocacy organizations. We can simply use the pillars to examine the choices we make and actions we take every day. The next chapter deals with some of the "big" attention-getting environmental issues that certainly deserve our consideration. But the big issues also tend to be the more difficult ones. The small individual issues are where we can have a more immediate impact. Remember that seemingly small actions can result in large impacts if enough people are involved. Also remember that these small issues and actions can deliver personal benefit to ourselves and enrich the quality of our lives. Let us consider a few examples, starting with the simple act of walking pictured in Figure 9.1.

Should I walk to work or school today? Let us put this question to the pillars. First, though we had better put the question into its proper context. This would be obvious to the person posing the question as they would know the circumstance and the alternatives but, for our example here, let us make it clear. Let us assume that our questioner is perfectly healthy and able to make this commute via foot and that the weather outside is not so frightful as to make the trip hazardous. Let us also assume that the distance of the commute is not too great. Now I realize that not too great is a relative measure, primarily depending upon one's age and just how much you enjoy walking. What we need here, to make this example realistic, is a distance where both walking and other alternate means of making the commute are essentially viable. A 15-mile (24 km) commute probably rules out walking for most of us. Let us pick a couple of miles (3.2 km) for the distance of the commute. This was my commute when I was in graduate school, and I walked it many times. If 2 miles (3.2 km) is out of the question for you, shorten it in your mind as you consider this example.

We also need to specifically identify the alternative to walking for this example. This is necessary so that we can properly score the impact of our proposed stroll against its alternative, its base case if you will. For our example, let us say that the base case, or alternative to walking, is driving your own car, which is a standard sedan powered by a gasoline engine. Under these assumptions, I would suggest the following scores (Table 9.1).

The environmental degradation pillar gets a score of +2 because we are emitting no pollution into the air like our car would. Even a car with a well-tuned engine and

FIGURE 9.1   Walking. (Illustrated by Mark Benesh. With permission.)

TABLE 9.1
Walk to Work/School Example

Pillar	Score
Environmental degradation	+2
Resource conservation	+2
Economic progress	+1
Personal benefit	+2
Total	+7
Indication	Hit the sidewalk!

pollution control devices on its exhaust will discharge some amount of carbon monoxide (CO), nitrogen oxides (NO_x), and hydrocarbons (HCs—molecules containing hydrogen and carbon atoms) into the air. This is in addition to carbon dioxide (CO_2) that is produced in significantly greater quantity than these other molecules. Carbon dioxide released through burning of fossil fuels such as gasoline is a concern with regard to global warming and global climate change. CO_2, however, is a direct result of the combustion process. It and water (H_2O) are formed when fuel (hydrocarbon molecules) is burned. It is a necessary and unavoidable product of combustion. If we want the energy to power our car, the CO_2 comes along for the ride. The other three, CO, NO_x, and HCs, are by-products that can be minimized with good engine design without negatively impacting the release of energy that we seek to keep our wheels turning. We will discuss the process of combustion and its relationship to global warming in more detail in Chapter 12. In any regard, driving our standard gasoline-powered car will emit air pollution, whereas walking will not. Even accounting for the carbon dioxide that we exhale with each breath, walking is superior to firing up our car's engine. I do not know about you, but even when I am driving my car, I am breathing, too. Our personal carbon dioxide emissions are relatively constant no matter whether we are walking or driving. Therefore, walking deserves the +2 score for the environmental degradation pillar.

Resource conservation is scored +2 because walking requires no fossil fuels to make our commute. As we learned in Chapter 6, actions or choices that do not consume fossil fuels will likely score a +2 on the resource conservation pillar when compared to actions that do consume fossil fuel. You might argue that walking does require us to consume other resources, more food for its calorific value, and perhaps more water to stay healthily hydrated. These additional resources that our bodies might require are insignificantly small, however, compared to the energy required to power our automobile. If you think about it, not only does our car consume fossil fuel where we do not, but it also requires much more raw energy to move it from point A to point B than it takes to move our bodies the same distance. A typical car that you would find on US roads weighs approximately 3000 pounds (1360 kg). Many large cars, trucks, and sport utility vehicles weigh more than that. I, on the other hand, weigh less than 200 pounds (91 kg). Driving my car instead of walking means I would have to move 15 times more mass over the 2-mile distance of my commute. That represents a lot more energy required. Think about the effort it would require if you had to push your vehicle the 2 miles (3.2 km) of the commute. All that energy, and more, is neatly stored within a gallon of gasoline.

Economic progress gets a +1 because the money we save on gasoline and auto upkeep more than offsets the wear and tear on our shoes; in other words, it is an economic gain for us individually. I was tempted to score this pillar +2, but there are some trade-offs to consider that kept me at the +1 score. The biggest trade-off is time, and if you believe the old saying, time is money. A 2-mile (3.2 km) commute on foot would take me about 30 minutes. In a car, it could take considerably less time than that, depending upon the traffic and the number of stoplights you encounter on your way. Let us assume that the commute by car takes 15 minutes. If you multiply that by two to account for both the morning and evening commute, then you would "save" a half hour every day by driving. This is a half hour every work or school day that you could invest in something else, something that could possibly have an economic gain associated with it. This time factor, depending upon how much value you credit it, could easily change the economic progress pillar score. If that half hour has a great value to you, I could see a score of zero or even −1 making sense. If that half hour merely represents one less half hour you spend at the gym, then a +2 score makes sense. Maybe it even saves you from having to keep a gym membership at all.

As you can see, this is a case where your own personal situation has a lot of bearing on the score for the economic progress pillar. I chose a +1 for this example because it made sense to me in my own situation, and because I thought it adequately represented a reasonable range of scores that I would expect to see from others who were making this judgment.

Personal benefit gets a +2 score for several reasons. First, a 2-mile (3.2 km) walk is a good way to get some needed exercise. There are not many of us that would not benefit from a regular dose of low-impact exercise, something to counteract all the time many of us spend sitting at our desks or workstations and not moving around much at all. Second, walking typically represents a relatively stress-free commute. I have heard of road rage, but I have never heard of sidewalk rage. You can let your mind wander on foot without getting into too much trouble. You do not have to pay attention to all the cars around you, or whether the green light ahead of you is about to turn yellow. You do not have to watch your speed or worry about the hidden cop waiting to give you a ticket. You can simply enjoy the sights and sounds you experience on your way as your footsteps predictably gobble up the distance between where you began and where you are going. The third reason for my +2 score on this pillar is that I simply enjoy walking. I like the feeling of mild exertion when I push my pace a little faster. I like the feeling of independence that comes from not having to depend upon a machine to get me where I need to go. I like not worrying about finding a parking spot, or if I need to stop and fill my car with gas.

The total score for this pillar analysis is +7! That is a very strong positive score. So everybody should lace up the sneakers and get walking, right? While there is some logic to the suggestion, this book is certainly not about prescribing an environmental ethic for everyone to follow. My purpose is rather the opposite in that the benefit of the pillars is in encouraging everyone to make up their own mind. I certainly hope that many would agree with the assessment in Table 9.1, but the point is to give people a good tool to logically analyze the situation for themselves. There would probably be cases where someone might disagree with my scores for each of the pillars. You might hate to walk or dislike braving the elements. You might possess a fine automobile that gives you so much pleasure that you want to drive it everywhere you go. I can understand that. I like cars, too, and I am a self-proclaimed environmentalist. You might quibble with my economic analysis or my evaluation of the pollution caused by driving the car. Use the pillars to best represent your own particular and individual situation, and my purpose will be served.

In this example, you can also see how important the assumptions and the particular situation are. If we compare walking with taking the bus or riding a bike, we could get a slightly different answer. The type of car we have could lead us to a different answer. Say, we have an electric car that we recharge at night, and that we have signed up for our local power company's "Save the Earth!" green power program where we pay a little more, but we are assured that the electricity we buy all comes from renewable sources, primarily windmills for our particular case. This changes the example so that the proposed action is to commute by foot versus a base case of driving our environmentally friendly electric car. Now our scorecard might look like this (Table 9.2).

TABLE 9.2
Walk to Work/School Example #2

Pillar	Score
Environmental degradation	0
Resource conservation	+1
Economic progress	+1
Personal benefit	−2
Total	0
Indication	Stay behind the wheel.

The environmental degradation pillar gets a score of zero because neither walking nor driving our electric car emits pollution into the air like a standard car would. The green power we buy to recharge our car's battery at night emits no air pollution, and thus, walking is no better than the alternative in this case.

Resource conservation gets a score of +1, because even though we use no fossil fuels in either case driving the car still requires much more energy than walking. If we do not consume the green power to recharge our car, then perhaps it would be available for some other use that might displace fossil fuel. Green power is a very good thing from a resource conservation perspective, but it is still not as good as not needing the energy in the first place.

Economic progress gets a +1 score because walking is still cheaper than driving our new high-tech car. The "fuel" in this case, windmill-generated electricity purchased from our local electric utility company, does have a cost to it. Driving our car incurs this cost where walking does not. Following the logic used in the previous example, if the time loss between walking and driving is significant to you, I could see a zero score for the economic progress pillar.

I scored the personal benefit pillar −2 in this case to represent the situation that we really like this new electric car. Let us assume that we bought this fancy environmentally advanced commuter vehicle specifically for this purpose. We bought the car, so we would not have to walk, or maybe so we would not have to feel guilty about not walking.

So our total score for this second example pitting our feet against our car is a very neutral zero. In keeping with a general tendency toward conservatism, negative and neutral scores tell us to reject the proposed action. In this case, it indicates that the best decision for us is to keep our right foot resting on the accelerator pedal. Let the green power stored inside our car propel us along our way.

Our walking versus driving examples produced different results primarily because we changed the assumptions. The pillars require that we compare our decision, or our proposed action, to some base case. The base case is often our present circumstance, the status quo. In our first example, the status quo was driving a traditional car. In the second example, we changed the status quo to driving an electric car. The status quo is a very important and powerful concept because it sets the stage and bounds the opportunity for employing the pillars. If the status quo is particularly awful, then it is pretty easy to think up a number of actions that would score well with the pillars. If you are a true-believer environmentalist already, then your status quo might be hard to improve.

The status quo is a result of many past decisions and choices that brought us to where we are right now. Our status quo is filled with a history of good intentions and imperfect information. There is nothing inherently wrong with it, but it often gets a bad rap. The activists of generally any persuasion typically argue against the status quo. That is why we have rallies, sit-ins, marches, and lawsuits. Protest is a movement against the status quo. While protest is usually well intended and is often socially necessary, it can goad us into hastily made decisions at times. Change for change's sake is not what the pillars are about, and certainly not what practical environmentalism is all about.

Let us consider another automobile example. (Can you tell that I really like cars?) In my experience, buying a car is one of the biggest purchasing decisions that I have made. It ranks behind buying a home, but it still represents a sizable expense. I purchased a new car in the year 2000. It was not an electric car back then as the technology was simply not available on a large scale. Even now as I write these words, the electric vehicle has not quite made the scene, but it sure looks like its day is coming soon. Instead, I purchased a Volkswagen Golf TDI four-door hatchback with a five-speed transmission. This is the same car I mentioned briefly in Chapter 6. It has a dark blue exterior and nice gray interior. It has power windows but unfortunately no sunroof—too bad. TDI stands for turbocharged direct injection and is used on diesel engines to provide decent performance and great fuel economy too.

At that time in my life I was in a state of transition. I was in the middle of a change of employment, as well as expecting a household move from one state to another. I also knew that it would not be a quick transition and that I would be traveling quite a bit for at least a year, maybe more. I expected to put a lot of highway miles on this car fairly quickly. The car that I was driving at the time was a two-generation family hand-me-down Chrysler. It was my grandfather's car, then my mother's, then mine. It was relatively old but had not been driven very many miles. It got me wherever I needed to go, but it was better suited to be an in-town commuter vehicle and did not fit my needs very well anymore.

So my current car did not really fit the definition of the status quo. Even though I was driving it at the time, it was not a viable alternative to my proposed purchase of the Volkswagen. A more appropriate status quo would be another new vehicle, perhaps one that was commonly purchased at the time. For argument's sake, let us say that my status quo was a common four-door sedan from a large manufacturer, nothing terribly flashy but something that sold well and was common for people to purchase. It could be a Chevy, Ford, or Chrysler if you prefer the American models; Toyota, Nissan, or Honda if you like Japanese cars; or even Volvo or BMW if you would rather have a European make.

To be honest, I did not apply the four pillars of practical environmentalism when I bought the car, at least not consciously. But, as you hopefully noticed in previous chapters, the four pillars often mirror some parts of our normal decision-making processes. During my search for a new car, I was definitely interested in a vehicle that was highly efficient in terms of fuel use. This desire, of course, is highly aligned

---

TABLE 9.3
Purchase New VW Golf TDI

Pillar	Score
Environmental degradation	+1
Resource conservation	+2
Economic progress	+1
Personal benefit	+2
Total	+6
Indication	Good choice!

---

with the resource conservation pillar. Let us use all the pillars and review my choice (Table 9.3).

Wow, I must have done all right! A +6 total score is a pretty strong indication that you are on the "right" track. Let us analyze each of the pillar scores to understand the logic behind them.

Environmental degradation gets a score of +1 mainly because of a reduction in air pollutants that follow from a reduction in fuel use. Burning gasoline, or diesel fuel as my Volkswagen does, in a car's engine does release some amount of pollution from the car's tailpipe and into the air. Since my Volkswagen is roughly twice as efficient on a miles-per-gallon basis as the status quo auto in this case, it produces roughly half the amount of pollution. This is not strictly true because of the different pollut- ants emitted by gasoline and diesel engines, but using half as much fuel for every mile driven gives the advantage to the diesel engine. Many might argue that diesel engines are "dirtier" than gasoline engines and point to older diesel trucks to empha- size their claim. My car is not a black soot–belching monster, however, and while it may emit different air pollutants than a gasoline engine, its great fuel economy gives it the edge overall. The reputation of diesels with regard to emissions does temper the environmental score though, so I chose a +1 to reflect this instead of a +2.

Resource conservation gets a score of +2, clearly. When I purchased my car, I knew of no other mass-produced reasonably priced vehicle that provided better fuel economy. Perhaps a moped would have done better, but that clearly did not fit my needs in an automobile. While my car does consume fossil fuel, so does the status quo car in my purchasing decision. My choice literally represented the most fuel- conscious car that I could pick at the time.

Economic progress receives a +1 score. The high-fuel economy, long-life engine, and relatively simple maintenance translate into low cost of ownership, which is definitely an economic benefit to me personally. The only economic drawback that I saw was that the car was slightly more expensive than if it came equipped with a gasoline engine that provided lesser fuel economy. The initially higher cost tempers the positive score for this pillar from a +2 to a +1.

The personal benefit pillar gets a +2 score. I bet you are not surprised. This was my intent after all when I started searching for a new car. I specifically wanted a highly efficient, high-miles-per-gallon automobile. It gives me pleasure to be able to drive nearly all day long on the highway without stopping for fuel. I love calculating my fuel economy and seeing it top 50 miles per gallon (21 km/L). I also like the gentle rumble that the diesel engine makes and the fairly decent acceleration that you get when the turbocharger kicks in. As I have admitted before, I am a bit of a car nut, so it should come as no great surprise that I own a car with a turbocharged engine. In this case, though the turbocharger improves performance while still providing excellent fuel economy. I tip my hat to the engineers in Germany that put this engine together. Granted I will not win many races with my car, but it is fun to drive, and I do not usually race anybody anyway.

In retrospect, I think I made a pretty good selection. Our simple pillar analysis proves this out as well with all four pillars reporting positive results. While I did not explicitly include consideration of environmental degradation at the time I bought the car, it is easy to see that it is not very difficult to include it in a formal decision-making process via the pillars. The human pillars of economic progress and personal benefit usually take care of themselves when we are faced with important purchasing decisions. Economics and personal desires often rule our decisions in fact. We may need to show some disciplined thinking to remember to include the "natural" pillars of environmental degradation and resource conservation in the mix. But that is the purpose of the pillars after all. They are the structure within practical environmentalism that helps us not forget that both the natural pillars and the human ones are important, especially if we are trying to live a more environmentally conscious lifestyle.

In buying my car I was strongly motivated toward resource conservation, and it showed in that both the resource conservation pillar and the personal benefit pillar scored the highest possible scores. It is not unusual to have the personal benefit pillar align strongly with one of the other pillars. That simply indicates that we are passionate about the action or decision that we are contemplating. An avid nature lover will likely see environmental degradation and personal benefit scores go hand in hand. An ambitious money hungry capitalist will likely have personal benefit and economic progress pillars that mirror one another. This is to be expected. Rational application of the method will, however, temper these potentially powerful couplings and hopefully encourage us toward balanced and thoughtful decisions.

Let us use the pillars for another simple example, perhaps one that might even seem to border on the mundane. Compact fluorescent light (CFL) bulbs are very much in fashion right now. They are seen as a very environmentally friendly product, primarily because they use much less energy than incandescent bulbs. Let us plug these modern marvels, similar to the one illustrated in Figure 9.2, into the scorecard against a status quo of incandescent light bulbs, the "normal old-fashioned" kind (Table 9.4).

The environmental degradation pillar gets a score of zero because you can argue that you are trading one type of pollution for another. The new bulbs require less electricity to use, and so there is less pollution generated by power plants to supply the electricity. There is, however, mercury present within the new bulbs, and mercury itself is considered a pollutant that must be dealt with during disposal of the bulbs. (If you are somewhat panicked by the idea of mercury in your home, you might have scored this pillar with a –1; your call.) I was tempted to score this pillar

FIGURE 9.2 Compact fluorescent light bulb. (Illustrated by Mark Benesh. With permission.)

TABLE 9.4
Compact Fluorescent Light Bulb Example

Pillar	Score
Environmental degradation	0
Resource conservation	+2
Economic progress	+2
Personal benefit	−1
Total	+3
Indication	Get the new bulbs.

a +1 because I think reduced air pollution from energy generation more than offsets the risk of additional mercury in the environment. An environmental degradation score of zero is pretty conservative I think.

Resource conservation gets a +2 because of the large reduction in electricity required. Most CFL light bulbs are much more efficient than standard incandescent bulbs in that they use approximately 75% less electricity per equivalent light output and lasts 10 times as long (Energy Star 2010). A 75% reduction in power consumption to provide the same or similar light output is a huge number! If you think about it in terms of driving your car, for example, this is analogous to quadrupling your fuel economy without impacting the power of your engine 1 bit. A 20-mile-per-gallon (mpg) V8 Ford Mustang suddenly becomes a very efficient high horsepower 80-mpg supercar. You could stomp on the accelerator and still feel good about your environmental ethics. The car that I have, the VW Golf TDI that I gushed about a few pages ago, would increase its fuel economy to a whopping 180 mpg as compared to its current 45 mpg or so. I could really go a long time between visits to the gas station then. Suffice it to, say, a 75% reduction in energy consumption is truly remarkable.

Economic progress gets a +2 because these new bulbs should save us money in the long run, both in the amount of energy they consume and in how often they must be replaced. The value of the electricity savings more than offsets the higher cost of the new bulbs and is significantly large relative to the purchase price of the bulb so that it represents a very high return on your investment. For example, the Energy Star Web site referenced earlier has a calculator that you can use to compare the economic impact of using CFLs. Here is a simplified version of the calculations.

A CFL uses 15 W of power to generate the same light output (lumens) as a 60-W incandescent bulb. The CFL should last for approximately 10,000 hours where the incandescent is expected to last for 1000 hours. Let us assume that we use this particular light for 3 hours every day and that our cost of electricity is 11.3 cents per kilowatt-hour (kWh). A kilowatt-hour is a measurement of energy and a standard unit of measure for residential electricity purchases. Note the difference in the units between power and energy. Power is an instantaneous measurement where energy is the amount of power delivered over time. Energy is power × time or watts × hours in this case. A kilowatt-hour is the amount of energy equivalent to using 1000 watts of power for 1 hour.

The cost of electricity for the CFL for 1 year is

15 W × 3 hours/day × 365 days/year × $0.113/kWh × kWh/1000 Wh = $1.86/year

The cost of electricity for a standard light bulb for 1 year is

60 W × 3 hours/day × 365 days/year × $0.113/kWh × kWh/1000 Wh = $7.42/year

If you add in the price of the bulb itself, $3.40 for a CFL and $0.60 for an incandescent, the total annual cost for a CFL is $5.26 per year, while the incandescent will cost you $8.02. The following table summarizes the costs for both bulbs (Table 9.5).

This simple analysis shows that you save $5.56 in energy per year ($7.42–$1.86) for spending an extra $2.80 ($3.40–$0.60) on the higher-priced bulb. The additional cost of the CFL pays for itself in 6 months and, at the end of a year, you have doubled your money! What bank can you go to where you can double your money in a year?

Table 9.5 also shows a total cost savings of $2.76 per year for buying and using the CFL bulb. These savings would be even greater in succeeding years as the CFL bulb is projected to last about 10 years under these conditions where the incandescent

TABLE 9.5
Light Bulb Comparison

	CFL Bulb	Incandescent Bulb
Power (watts)	15	60
Energy (Wh per year)	16,425	65,700
Energy cost ($/year)	1.86	7.42
Purchase price ($/bulb)	3.40	0.60
Total cost ($/year)	5.26	8.02

would need to be replaced every year. Certainly, the absolute savings in this example are relatively small, just a few bucks, but the percentage savings and return on investment are relatively huge. Perhaps you have heard the saying "It takes money to make money." That old saying usually implies that it takes a lot of money to make a large return on your investment. Here, though, is an "environmental action" that requires little money to generate a great return on investment. It is an ideal case for the individual wanting to make a positive environmental impact. That is a +2 score on the economic progress pillar in my book.

Personal benefit gets a −1 in this case because maybe you just do not like the light quality as well, and these new bulbs are really finicky when they get cold. I personally would score this pillar a zero, but there are lots of people who seem to notice a difference in the light CFLs produce, so let us stick with a −1 and be somewhat conservative.

So our total score is +3, indicating that the best decision for us in this case is to go ahead and get the new bulbs. Again, it is good to note that if you scored things slightly differently, you could have easily arrived at a total of +1 or +2, still a positive result though prodding us to take action. If you came to visit me in my home, you would see these CFLs all over our house. My family has gotten used to the difference in light quality, and we have found that they do work outside, even on a cold Iowa winter night. They are not very bright when first turned on when it is cold, but they provide the light we need. I try not to stay outside too long on a cold winter night anyway . . . brrr!

This light bulb example is also interesting in that it demonstrates a case in which the personal benefit pillar is at odds with the total pillar score and final recommendation. This is the only example in this chapter that highlights this condition. The +2 scores of the resource conservation and economic progress pillars ultimately drove the decision in the positive direction for this example. This is exactly what we hope for when we apply the pillars. We hope that we can rationally analyze each pillar and allow ourselves to consider an action that might run contrary to our personal preference. We do not deny personal preference, of course. It has its place, and the personal benefit pillar exists to grant its due consideration. Practical environmentalism hopes to open our minds to realize the benefits of doing something that we might not have initially wanted to do.

This might seem like a lot of effort to go through to make a simple choice where we might have known the right answer from the very beginning, or thought we knew the right answer from the very beginning. These examples were intended to be instructional, of course, and hopefully illustrative of the method and logic of the pillars. The intent of the pillars is to force us to examine our own values and motivation in a realistic and holistic way. I also hope that it shows that seemingly simple decisions may be worthy of a bit more reflection and that the conclusion you reach can depend greatly upon your personal status quo which could very well be different from somebody else's status quo.

Now you might also ask, "Why bother with the pillars if there is not a "right" answer when you are done?" Good question. You can get lots of different answers, depending upon your own personal values and the assumptions you make when arriving at the scores. The point of practical environmentalism is really the execution of the method rather than the discovery of the answer. Not arriving at the "right" answer is not a problem because there is no such thing as the "right" answer anyway. The answers that make sense are those that move us, individually and collectively, in the "right" direction. The "right" direction is progress.

Notice, I did not say environmental progress. I wanted to say it, as I think it a worthy goal. But the pillars are not about environmental progress in absolute terms. The pillars are about environmental progress in context. The pillars will help us place the environment in its proper context as one of the four considerations we must make if we want to make sustainable choices and sustainable changes.

Now, I have uttered the "S" word, "sustainable!" And, I have even said it twice in one sentence. I have tried hard not to say it up to this point. It is a tough word because it can mean so many different things, and it is so easily overused. I do not want to turn this book into a sermon on sustainability. That is a quagmire I would prefer to avoid. Let me give it just a few paragraphs and we will move on.

Sustainability is a great concept if you do not look at it too critically. It basically means that we should leave things at least as good as we found them. In other words, do no harm. It is tough to argue with the premise. The devil though is truly in the details. Who defines the qualitative "good" and "harm?" What is the timescale over which we should consider the implications of our decisions? And, does sustainability deliver any benefit to individuals? Let me illustrate with a short story.

I used to work in a power plant. We turned coal into electricity for a small midwestern city. I worked with an instrument and control technician named Dan who usually had something interesting to say. I made a point of listening to him because he was pretty good at his work, and his work was really important. Instruments and controls are the lifeblood of a modern day power plant. If you do not do that well you are bound to end up in the dark.

Dan introduced me to the term *battle short*. It is a military term and describes a special switch that can override normal protective relays and safeguards in order to get equipment to run in an emergency. Dan's experience with the battle short switch was with engine generator sets that were used to make electricity to power all sorts of equipment. These engine generator sets are expensive and often critical to a unit's mission. Modern weaponry has become very sophisticated and somewhat hungry for electricity. No longer does a soldier merely shoot at his enemy with a rifle. A modern army has advanced communications gear, remote-controlled surveillance capability, and weaponry that can be guided to its target. Computers are equally at home on the battlefield as on the desk in your cubicle. All this modern military equipment requires electricity to operate, and there are not very many electrical outlets in the sands of Iraq or in the mountains of Afghanistan. The old saying that an army travels on its stomach is probably still true, but a modern army must bring their electricity generators along with them too.

Dan told me the following story of working on engine generator sets when he was stationed in Germany. He tells it in a much funnier fashion than I can relate here; storytelling is one of his gifts. Any errors in the retelling are mine.

Dan happened to be working on a base that was preparing to participate in a missile simulation exercise. The area where he was contained a set of six high-frequency generators, 400-cycles-per-second machines. As the test exercise began, Dan hung around in the control room to observe the action. He watched as five of the six generators came on line and started generating power. The sixth, however, was not cooperating, and the soldier operating the units was not having any success figuring out what was wrong. Dan watched as this situation drew the attention of a lieutenant, then a captain, and finally a major. The enlisted man trying to start the troublesome sixth generator became paler and paler as the number and rank of officers increased. He glanced back at Dan occasionally, knowing Dan was a technician with some experience on these units. However, it was not Dan's place to insert himself into another company's operation, and he remained silent in the background. Finally, one of the officers noticed the generator operator's frequent backward glances and asked him why he kept looking at Dan. The operator explained that Dan was a technician for these units, and Dan was immediately summoned by the officer.

"Can you get this generator running?" the officer asked.

"Yes, sir," Dan replied.

"What is wrong with it?" The officer continued.

"I do not know exactly sir," Dan replied. "It is probably in one of the safety circuits, oil pressure, or the electrical system maybe."

Engine generator sets, as well as many other complex and expensive pieces of military equipment, have safety systems built into their controls. The idea is to make them "foolproof" and not let them start unless all the auxiliary systems are able to support the start-up and continued operation. You want to make sure that there is an adequate supply of lubricating oil, for example, or that there is sufficient fuel pressure to fire the unit. These safety systems are included to make it more difficult to ruin the equipment.

"How can you get it going then?" the officer asked, probably knowing that fixing such a problem could take some time, and he did not have much time to get the power he needed to play his part in the battle simulation.

Dan responded, "Just flip the battle short switch and fire it up, sir."

"Do it then," the officer commanded.

"I cannot, sir, not without a written order," Dan replied. "It will probably ruin the generator."

Dan did not want to be tagged with ruining a generator. He had known other soldiers who screwed up in one way or another and ended up serving an extra tour of duty. Dan was not at all interested in that.

"I am giving you a direct order in front of all these witnesses," growled the frustrated officer. "Start the unit!"

"Yes, sir, that is all I need," Dan said, knowing that direct orders, from an officer and in front of witnesses was something a soldier does not object to under just about any circumstance.

Dan instructed the generator operator to flip the battle short switch and start the unit. It worked. They brought the sixth generator on line and were able to run the missile simulation and successfully take part in the battle exercise. As Dan predicted, the generator soon seized up and quit running, but it had served its purpose. Now it would be up to the mechanics to rebuild it if they could.

You can, of course, criticize the commander's judgment and wonder why you would want to damage a pricey piece of equipment for sake of a display, but the battle short feature worked as designed. It circumvented the safeguards and let the generator run for a brief time. The intent of the battle short circuitry is really nobler than that as it is meant for battle conditions where life and death are in the balance. The life of a generator is no match for the life of you and your comrades. When your position is about to be overrun and your buddies are dying all around you, you flip the switch without a second thought.

So here is the analogy to sustainability. The battle short switch is akin to environmental damage, the opposite of sustainability. It is inadvisable in most instances but necessary in some. The sustainable path is, of course, to avoid the switch, but sustainability gets trumped by our survival instinct. Sustainability can also be superseded by our instincts for wealth, power, prestige, promotion, and all sorts of other short-term human emotions. The commander's life certainly was not in danger when he instructed Dan to use the battle short switch, but perhaps he felt his career was in jeopardy if he did not.

Part of the problem with the concept of sustainability is that there are few personal benefits for practicing sustainability. To the contrary, the personal benefits usually come with the battle short switch. The benefits of sustainability tend to accrue over the longer term to the generations that come after us. Therefore, sustainability is much more important to societies than to individuals. On the individual level, it tends to be very abstract, difficult to measure, and really almost meaningless.

I do not intend to discourage sustainable thinking and sustainable actions. I do think it is important. Sustainability is not easy though and not the ideal framework to use if you want to inspire meaningful environmental progress based on individual actions and choices. The pillars are better suited to this end as they are more direct, easier to incorporate into our daily lives, and can provide immediate personal benefit. And let us not kid ourselves into thinking that personal benefit is not important, or that somehow we can tap the most noble and honorable parts of the human psyche and start living our lives for the benefit of the environment instead of for ourselves. That is a pipe dream. Personal benefit is crucially important. At the risk of repeating Chapter 7, personal benefit is what drives so many of our decisions. Our task is to identify actions where we can align all or most of the pillars, including the personal benefit pillar if possible, in a positive direction.

There is a way, however, in which we may consider sustainability within the concepts of practical environmentalism. Sustainability within this context could be shown as the general alignment of all four pillars in the positive direction. This is slightly broader than the typical definitions for environmental sustainability as it includes the crucial personal benefit pillar. Practical environmentalism asserts that the individual is important, and that true sustainability incorporates the environment and its resources along with our personal economic situation and our set of values and preferences.

The pillars are also intended to make it more difficult for us to make mistakes. They are just as valuable in promoting environmental benefit as they are in resisting ill-founded environmental actions or policies that come at too high a cost in terms of the other three pillars. The pillars encourage us to make choices that are generally beneficial and that have a low risk of undesirable surprises. They prevent us from blindly following the environmental crisis of the day without first thinking of the impacts of our actions. The pillars are meant to help us realize steady environmental progress, sustainable progress if you will.

I suggest that the sustainable progress that practical environmentalism calls for starts small. It begins at the individual level with actions and choices in which we personally have much control over the outcome. Indeed that has been the focus of this chapter and the examples we discussed. We can apply the pillars and seek out environmental betterment through our daily choices of how we will commute to work or school, what kind of car should I own, or which type of light bulb will I use in my home. Of course, there are many, many other choices we make in our personal lives that impact the natural world and where we could easily apply the practical environmentalism pillars. The food we eat, the clothes we wear, the products we buy, and even the way we spend our free time are all candidates for thoughtful reflection on the environmental impacts of our actions.

This is indeed the place to begin. Before we become too concerned with acid rain, global warming, nuclear waste, or ozone depletion, let us make sure that our own personal choices reflect our own personal environmental ethics. It is too easy to complain about what others are doing or what they should be doing and are not. Practical environmentalism has something to say about these big issues too, but it is more relevant if we focus on our own actions first, because in reality many of our actions are creating these big issues in the first place. Practical environmentalism is very comfortable in the grassroots approach to environmental progress. Not only does it realize that the positive action of many individuals can represent significant environmental progress but also that as we are individually willing to test our own personal choices through the practical environmentalism framework, we will become better able to respond to the daunting environmental issues that we face. In the following chapter we will begin to discuss these big issues, the ones that we hear so much about, but that seem beyond our grasp and influence.

REFERENCES

Energy Star. 2010. Light Bulbs (CFLs). http://www.energystar.gov/index.cfm?fuseaction=find_a_product.showProductGroup&pgw_code=LB (accessed October 6, 2010).

10 The Pillars and the Really Big Issues

Up to this point we have discussed the rhyme and reason of the pillars. We now know what they mean and how to use them in our daily lives. However, do our daily decisions make any difference with regard to all the big environmental problems that are constantly brought to our attention? Can the pillars, and practical environmentalism in general, be applied to the really big issues of the day?

My answer to that question is, "YES," of course. I do not imagine you would have expected anything else since I went to all the trouble to write this book in the first place. My "YES" is qualified, however. As you know, the pillars are intended to be individually useful, and individually practical. Therefore, to use the pillars with the really big issues, we must frame them in such a way that the scorecard is meaningful to us individually. To say that "Water pollution is bad" or "We must act quickly to stop global warming" is useless as far as the pillars go. We must frame the question so that we have a proposed action to compare against the status quo and that the result grants us valuable insight that we can act upon.

We will look at several examples to learn how the pillars can be applied to the really big issues. First, though, we should examine their qualities and characteristics in general, as they tend to be significantly different from our personal environmental choices.

First, and by definition, the really big issues are really big! They cannot be "solved" by you or me. They might not even be issues to you or me. We may not care one bit about them. However, a number of people do care, and a sufficient number have a big enough voice to make these issues heard by the rest of us. The big issues are those that expand across space and time to affect other people, towns, states, countries, and even generations. These are issues such as air pollution, water pollution, species extinction, ozone holes, invasive species, genetic mutation, and hazardous waste, among others. The biggest issue of our time, the granddaddy of them all, as the famous college football and Rose Bowl announcer Keith Jackson might say, is climate change, also known as global warming, as it receives the most attention by far. We will get to that one in Chapter 12.

Second, the really big issues grow to have a life of their own. Initially championed by a relatively small group, be it scientists, politicians, activists, or some combination thereof, the really big issues become incarnated into our culture and generally defined as a problem that needs to be fixed. The media are often complicit in the evolution of an environmental concern into a really big environmental issue. It is after all their role to bring these things to our attention, but there is a fine line between providing information and competing for ratings. We ourselves are also complicit whenever we accept the teaser headlines and sensationalist propaganda as truth instead of realizing they are, essentially, a marketing ploy.

The far majority of the environmental information/propaganda an individual receives comes from the media, commercials, and news in particular. We generally find little time to do our own individual research on these issues, and so we find ourselves trusting in what we are told. Since the really big issues have grown to be above reproach, the media can reiterate them at will with very little thought or justification. They tell us what we want to hear, or expect to hear, and grab some ratings in the process. As in the old fable, the naked emperor wore a fine set of clothes.

When an issue becomes really, really big, it becomes ingrained within our corporations. That is how you know that an issue has truly made it to the big time. Corporations now use valuable commercial time and advertising space to convince us of their environmentally responsible ideals instead of promoting a good or service they provide. It has become very trendy to be green, or at least to appear green. The reason for all these advertisements, of course, is to gain a marketing advantage by being seen as in tune with the consumer. It makes business sense.

To argue against these really big issues or support responses that do not address an issue in its entirety becomes unpatriotic, immoral, or worse. The big issues often develop an inertia that is hard to contain regardless of pesky facts or new information. The big issues and our mindless cultural acceptance can run right over individuals who bother to question them.

Finally, the really big issues are complex, costly, difficult, and fraught with uncertainty. There are no easy answers. Proposed "solutions" tend to be costly, too, and may offer little guarantee of success. We often hear about the dire ramifications of the really big issues, but we hardly ever hear of the risks of the proposed solutions. This would not be so troubling except that "solutions" are really big and costly, and wasting a lot of money and resources on an ill-fated solution prevents us from making real environmental progress by investing in a not-so-big issue that is much easier to address.

So the primary benefit of the pillars with respect to the really big issues is to evaluate proposed "solutions." The pillars can easily serve as a basic sanity check by answering the simple question, "Would you or I support the proposed solution?" We can do this by running the solution through our own personal scorecard to see whether we get a positive result. Let us look at a few historical examples of big environmental issues and apply our own personal scorecard to see whether we would have agreed with the chosen action at the time of its decision and implementation. Let us begin by returning to the example of the construction of the Hetch Hetchy Dam that was discussed in Chapter 2. If you or I were alive in the early 1900s, would we have supported this massive government project? Let us find out.

Recall that the construction of the Hetch Hetchy Dam was a monumental engineering project in its day, and most notably was a major economic development project within a US national park. It flooded a pristine and extraordinarily beautiful mountain valley within the Yosemite National Park in California. It inspired a ferocious debate pitting the city of San Francisco against the Sierra Club. There was a very high degree of political posturing and maneuvering that lasted for years and was complicated by the great San Francisco earthquake and fire in 1906. Ultimately, the dam was built at great cost both in monetary terms and in lives lost during construction.

TABLE 10.1
Build the Hetch Hetchy Dam

Pillar	Score
Environmental degradation	−2
Resource conservation	+2
Economic progress	+1
Personal benefit	0
Total	+1
Indication	Build the dam

Let us assume that you or I were alive back in the first decade of the twentieth century and that we were aware and interested in this potential project. Let us also assume that we were "privileged" with the knowledge of our modern environmentalism and were not constrained by the conservation–preservation environmental dichotomy that existed 100 years ago. Granted this perspective is not historically appropriate, but hopefully it will be instructive in applying the practical environmentalism framework to the proposed dam and seeing if we would now agree that it was a worthwhile endeavor. The status quo for comparison in this case would be literally to do nothing. Let the Tuolumne River valley remain as it was, not completely undisturbed but at least above water. Here is what my pillar scorecard would look like (Table 10.1).

Environmental degradation gets the most negative score possible. A local ecosystem was destroyed and replaced by another. I could easily argue for a less negative score here, but I chose the −2 score to be as "environmentally friendly" as possible for the sake of this example. There are some environmental benefits that accrue due to the emission-free electricity produced by the dam that could merit a −1 score. Your own personal ethics and values are of course very influential in deciding which of these negative scores is most appropriate.

Resource conservation, on the other hand, gets the most positive score possible in my analysis. The dam created a very large reservoir that allowed San Francisco and surrounding areas more control in managing their water supply. The dam also created a source of hydroelectric power and generated approximately 500 MW of electricity. That amount of electric power is equivalent to a medium-size coal-fired power plant today but without the air pollution associated with burning coal. Hydroelectric power is renewable and controllable and tends to conserve fossil fuel resources. Then and now we humans have a need for water and electricity. We relax when these needs are in plentiful supply, and we get very nervous in times of scarcity.

Economic progress is scored positively at +1. Jobs were created during the construction and operation of the dam, and it provides a useful commodity in water and electricity. Naturally we find fault today with the early 1900s construction practices that allowed workers' lives to be forfeited in the name of progress. This category could easily have been scored a +2, but I chose to restrain it to +1, again in light of being "environmentally liberal." I would find it difficult to score this pillar a zero, but I could easily see someone scoring it a +2.

I scored the personal benefit pillar a zero. There are a number of ways to approach this pillar and arrive at this score. First, if we imagine ourselves residents of the east coast of the United States, or perhaps a midwestern city, then the Hetch Hetchy issue in the mountains in eastern California might not affect us personally in any way, prompting the neutral score. Or if we imagine ourselves a resident of California, we might rate the damming proposal a −2 if we valued a relatively undisturbed mountain valley and would miss the scenic hikes along the valley floor. We might also rate this a +2 if we valued the beauty of the newly created lake amid the mountains. I decided the neutral score would represent the disinterested easterner or an average composite of Californians.

So, in general, I could see myself cautiously supporting the project. Certainly the personal benefit pillar plays a large role in determining the overall result, and it is easy to see how a passionate environmentalist could justify his or her opposition to the dam. I think the zero score for this pillar is relatively fair though in describing what I would expect from a majority of potential reviewers of the issue. Again, recalling from Chapter 2, this issue was ultimately settled in the political arena, so examining the values of the population is a worthwhile exercise in examining the level of support for the project.

An interesting twist to the Hetch Hetchy saga is the modern notion of restoring the valley to its pre-dam condition. This proposal is to literally remove the dam, drain the reservoir, and allow the Tuolumne River to return to its "normal" historical course. This proposal continues to receive support among environmental activist organizations (Sierra Club 2004) and has even sparked some interest within government from time to time (State of California 2006). Let us also apply the pillars to this modern twist on the historical struggle (Table 10.2).

Environmental degradation gets a +1 score to account for the righting of a historical wrong from the friendly environmentalist's perspective. Note that I did not score it a +2 as you might expect, that being the direct opposite of the −2 score in the previous example. This is primarily due to my own level of uncertainty around the ability of the flooded ecosystem to return to its pre-dam state. Would we really see the valley return to what it once was, or would it be a muddy adulterated mess? I chose the +1 to represent the expectation that the valley would eventually return

TABLE 10.2

Remove the Hetch Hetchy Dam

Pillar	Score
Environmental degradation	+1
Resource conservation	−2
Economic progress	−1
Personal benefit	0
Total	−2
Indication	Keep the dam

to some semblance of its former state and that would be an environmental benefit in general. Another item to keep in mind is the loss of the pollution-free electricity produced by the dam, and you could argue that this represents some level of environmental degradation and therefore justifies a +1 score for this pillar instead of a +2 score.

Resource conservation, on the other hand, gets a −2 score in my analysis and is the direct opposite of the score in the previous example. What was gained in water supply and renewable electricity production would be lost. Five hundred megawatts of electricity production would have to be replaced. Assuming renewable forms of electricity generation are already committed and dispatched to the electrical distribution grid, this 500 MW of power would likely come from fossil fuel sources.

Economic progress is scored negatively at −1. Certainly jobs would be created during the demolition of the dam and the restoration of the valley, but the money expended for these activities would not provide an easily quantifiable economic commodity such as water and electricity. This proposal is similar to an environmental cleanup where costs are high and benefits to the economy itself are low.

I scored the personal benefit pillar a zero for similar reasons as in the first example. Many of us have no personal attachment to this particular location, and of those who do, some could conceivably find value in the existing reservoir while others might prefer the more natural state.

So the total score of this pillar analysis is −2, indicating that we should keep the existing dam. As you might expect, this is very much in line with the first example examining the dam's construction. With relatively little effort we were able to analyze a fairly big environmental issue and rationalize and quantify our support for the dam that flooded the beautiful Hetch Hetchy valley. The pillars showed this historical example to be neither a clear winner nor a clear loser. Final scores of +1 and −2 in these two examples are close enough to zero to show this dam to be a potentially contentious issue, just as its history has showed it to be.

Let us now consider another historical example, but a more recent one. During the mid-1980s we became aware of something called the ozone hole. It was discovered by British scientists working in the Antarctic (NSF 2010). Most of us had no idea what ozone was, but we gradually became educated. We learned that ozone is essentially an oxygen molecule with an extra oxygen atom. Whereas the oxygen we breathe, O_2, has two oxygen atoms, ozone has three, O_3. Ozone exists in small quantities at different levels within our atmosphere. The scientists tell us that ozone in the stratosphere serves a very important function of filtering harmful radiation from the sun before it reaches the earth's surface. Without ozone we would be more susceptible to skin cancer, cataracts, and suppressed immune response, and many plants and small organisms could be damaged as well. Ozone floating around high up in the atmosphere protects us against these dangers without most of us having any idea what is going on.

British scientists discovered that there was an area in the stratosphere above the South Pole where ozone was severely depleted during the winter months. This became known as the ozone hole. We were told that the situation would get worse with time and that we humans were causing the problem by releasing certain chemicals into the air that destroy ozone (Rowland and Molina 1974). If we did not do something to fix this problem, the ozone hole might grow bigger and affect more geographic regions than just the South Pole.

Let us assume for our purposes that all this information was true. It would be worthwhile for these claims and theories to be tested and proved, but that would involve a level of scientific detail and expertise that is beyond the scope of this book. I certainly encourage a healthy dose of distrust when it comes to claims of environmental calamity, but even lacking complete and sure knowledge of the environmental dangers, we can employ the pillars to gauge and direct our actions. For this example, we will trust that the appropriate scientific review was done and that there are no major flaws in the scientists' theories.

To prevent the feared environmental damage, we were told that we must stop using certain chemicals, chlorofluorocarbons (CFCs) to be exact, which were found in aerosol products and refrigerants. At the time, there were no readily available substitutes for these products, so there were no immediate easy answers. Eventually, a consensus emerged among many of the industrialized nations that banning these CFCs was necessary and should be phased in over a few years to allow time for the development of alternative products. This consensus was formalized in an agreement known as the Montreal Protocol. It led many nations to ban the production and import of CFCs and caused industry to develop and use alternative products that were significantly more expensive than CFCs. These additional costs were passed on to consumers in large part.

So, was this decision appropriate to the environmental issue? Would you or I have supported the ban at the time? Let us apply the pillars and see what we come up with (Table 10.3).

Environmental degradation gets a +1 score in my analysis, primarily based on a belief that banning CFCs will result in an environmental improvement and increase our chances of avoiding future environmental distress from a lack of protective ozone in the stratosphere. Recognizing that some of the confounding factors often present in environmental decision making (uncertainty, measures of success, fallacy of prediction, assumption of future states, etc.) mentioned in Chapter 4 could apply here, I hesitate to give a +2 score for environmental degradation.

Resource conservation is scored a zero in this analysis. I do not see any obviously large impacts to the use of natural resources. Banning CFCs will result in the development of other products to take their place. Without a very detailed analysis, it is hard

TABLE 10.3
Ban Chlorofluorocarbons (CFCs)

Pillar	Score
Environmental degradation	+1
Resource conservation	0
Economic progress	−1
Personal benefit	+1
Total score	+1
Indication	Bye-bye CFCs!

to know whether the manufacture and use of these new products will result in more or less natural resources being consumed. In this situation I feel a score of zero is appropriate.

Economic progress gets a −1 score. It is pretty clear that the replacement refrigerants will be more expensive and that we will personally bear the additional cost whenever we buy or service air conditioning systems for our homes or cars. I do not expect these costs to be so huge that we would not be able to afford air conditioning, though so a −2 score is not warranted.

Personal benefit gets a +1 score in our analysis. If the scientists' theory is correct, then banning CFCs will lower our individual risk for some diseases. That is one less thing for us to worry about and certainly a benefit. Antarctica is an awful long way away though, and I do not plan to go there anytime soon, so a +2 score seems unjustifiably high.

Thus, our total tally with the pillars is +1. A negative economic score is slightly overtaken by positive environmental and personal benefit scores. Banning CFCs seems to be a worthwhile activity, and this pillar analysis supports our government signing the Montreal Protocol.

Let us examine another really big issue—acid rain. In 1990 the United States amended the Clean Air Act to institute a program to reduce acid rain. Acid rain, as its name implies, is rain or precipitation in general that is more acidic than it should be. Over the years, rainfall has become more acidic because of the increased presence of sulfur dioxide (SO_2) and nitrous oxides (NO_x) in the atmosphere. These airborne chemicals come primarily from electric power plant emissions. These emissions react with water and particulates in the air and acidify the water vapor. The water vapor forms clouds and eventually returns to the earth as precipitation. Acid rain has been blamed for killing fish in lakes, damaging building concrete and stonework, and degrading air quality, which negatively impacts human health.

The solution to this problem, incorporated into the Clean Air Act amendments, was to regulate the amount of SO_2 and NO_x produced from power plants. This regulation took the form of a cap-and-trade program for SO_2 and more stringent emission limits for NO_x. The SO_2 cap-and-trade program restricted the total amount of SO_2 that could be emitted into the air, but it did not limit the amount that could be emitted from any individual site. Instead, existing sites were given an allowance for a relatively small amount of SO_2 emissions and would have to purchase additional allowances from other sites if their original allowances were insufficient to cover their emissions. This created a market for SO_2 allowances, which is the "trade" component of the cap-and-trade program. Now individual sites had a choice in complying with the new acid rain regulation. They could reduce their emissions to a level that would not exceed their allowances, or they could purchase additional allowances to cover their excess emissions. This choice could be quantified in economic terms, and emitters could calculate the cost to comply with the new rules. Some power plants might install emissions control equipment to reduce their SO_2 emissions, some might change the type of fuel they use to one containing less sulfur, some might shut down entirely, and some might do nothing but purchase additional allowances.

So, was cap-and-trade a good approach to reduce acid rain? Can we support a market-based solution to a pollution problem? Can we let some polluters off the hook

TABLE 10.4

Cap-and-Trade Sulfur Dioxide Emissions

Pillar	Score
Environmental degradation	+2
Resource conservation	−1/0
Economic progress	−1/0
Personal benefit	+1
Total	+2
Indication	Bye-bye sulfur

by not requiring them to reduce an individual plant's pollution? Let us use the pillars and find out (Table 10.4).

Environmental degradation gets a +2 score in the analysis. Here we have a case of a pollutant causing direct and measurable harm. We do not have to worry too much about the confounding factors of uncertainty and fallacy of prediction as the damage has already occurred. It is measurable, and the scientific theory and evidence linking SO_2 and NO_x emissions to acidic precipitation is strong. While cap-and-trade does not require specific reductions at every power plant, it does mandate a significant national reduction in SO_2 emissions, and we consider that to be a definite improvement. Also recognizing that cap-and-trade does not allow power plants to increase their SO_2 emissions above their current permit limits, we should not expect any deterioration in local air quality due to a particular plant choosing to purchase allowances instead of reducing emissions. This strongly positive score indicates a belief that cap-and-trade will result in a marked environmental improvement.

Resource conservation is scored somewhere between a −1 and zero in this analysis. Admittedly, we are splitting hairs here, but either score could probably be justified depending upon the assumptions. Cap-and-trade will likely shift the balance in the types of fuels used. High-sulfur fuels (eastern coal, heavy oil) will likely be used less, and low-sulfur fuels (western coal, natural gas) will likely be used more. Cap-and-trade will also likely require the installation of some additional pollution control, which will require power and energy to operate and will thereby consume more resources. Cap-and-trade could encourage the shutdown of older, less efficient power plants, which would likely result in a net benefit in resource conservation. Again, without a detailed analysis, it is hard to know the exact magnitude of the impact to natural resources, but I expect it to be slightly negative. I chose to represent this expectation with a modified score of −1/0.

Economic progress also gets this hairsplitting −1/0 score. The case here is akin to the resource conservation situation in that some actions under cap-and-trade will likely result in economic efficiencies, while others will likely impose economic penalties. Any additional costs will likely show up in higher electricity rates and be imposed upon consumers—us. By the time these costs are spread across all the electric ratepayers, I expect the increase to any individual consumer to be relatively small. Most of us probably would not even notice it on our bill. So I expect a slightly negative economic impact but one that is probably not worth a −1 score. I will represent this again with a modified score of −1/0.

Personal benefit gets a +1 score in the analysis. Reduced levels of acidity in precipitation and decreased amounts of SO_2 and NO_x in the air we breathe are valuable to us as individuals. While meaningful, I really do not expect the benefits to be extremely large. If you are particularly interested in or affected by air quality, I could easily see you rating this pillar a +2.

So, our total tally with the pillars is +2. We count the −1/0 scores for resource conservation and economic progress as minus one-half each. These mildly negative scores are overshadowed by positive environmental and personal benefit scores. Our pillar analysis shows cap-and-trade to be a worthwhile program and rationalizes our support for this part of the 1990 Clean Air Act amendments.

These last two examples are widely regarded as environmental success stories. The Montreal Protocol is noted for its high level of international cooperation, and cap-and-trade seems to have delivered real reductions in emissions and acid rain. Broad cooperation, even international cooperation, is widely seen as a huge benefit, if not outright necessity, in addressing the really big environmental issues. Cap-and-trade as a market-based approach to solving big environmental issues is somewhat more contentious. "Hard-core" environmentalists might prefer a more prescriptive approach where government sets individual emission limits and industry is forced to comply under penalty of law. "Industrialists" probably appreciate the flexibility of the market-based approach and feel that they are better able to manage their enterprise in this environment. The debate between prescriptive and market-based approaches is certainly not over.

In both examples, the pillars indicated that the proposed action was worthwhile. At the individual level, the proposed action was beneficial to us and we could accept it. Now, in reality, our government probably would not solicit our opinion on these issues. We would have to go to some effort to be heard. We would have to call our congressman or call the USEPA or make a fuss in front of the local media in order to have our individual opinion noticed. In all likelihood most of us would not bother, especially since we think the government's action is OK in these cases.

Let us look at one more example, one where government regulation was applied directly to a resource conservation issue. In 1974, speed limits across the United States suddenly became constrained to a maximum of 55 mph (89 km/h). This was thanks to the Emergency Highway Energy Conservation Act, passed by the US Congress and signed by President Nixon (USDOT 2010). This act itself did not mandate that we slow down to 55 mph, but rather withheld highway funding from any state that did not set and enforce the 55 mph limit. Federal funding is the lifeblood of state highway construction departments, and the act quickly achieved 100% compliance from the states. Figure 10.1 shows the dreaded sign that sprouted up on every major high-speed road in the nation.

The federal government applied these heavy-handed tactics as a reaction to the Arab oil embargo of 1973 and the gas shortages and economic disruption that followed. It was perceived to be a matter of national security. Many Arab nations were displeased with United States support of Israel at the time and used their leverage over oil production to put political pressure on the United States as well as the North

FIGURE 10.1 US national 55 mph speed limit. (Illustrated by Mark Benesh. With permission.)

Atlantic Treaty Organization (NATO) member countries. Gasoline prices climbed as supplies were curtailed.

The Emergency Highway Energy Conservation Act specifically targeted resource conservation, but from a political and national security perspective instead of an environmental perspective. This was obviously not an action born out of a desire to improve the environment. It simply was a supply-line issue. There was plenty of oil and gas available, just not in the United States. Both in the United States and around the world, gasoline became more expensive. Notice that the name of the act includes the word "Emergency." It was obviously perceived as a crisis.

Often, crises produce hasty and imperfect decisions. That is to be expected. I imagine most American drivers were not very happy with the shortage of gasoline, nor the new restrictions on how fast they could drive down the highway. However, the act originated within the US Congress, which is elected by the citizen voters of the United States, so it is fair to assume that our congressional representatives felt their constituents would at least grudgingly support the act. Again, it was a matter of national security.

TABLE 10.5
Institute 55 mph Speed Limit

Pillar	Score
Environmental degradation	+1
Resource conservation	+1
Economic progress	0/+1
Personal benefit	−2
Total	0/+1
Indication	Slower is better?

Let us look back on the Emergency Highway Energy Conservation Act and apply the logic of practical environmentalism to it. Let us see if we would agree that it made sense from an individual's perspective (Table 10.5).

Environmental degradation and resource conservation pillars both get a score of +1 for the basic reason that driving slower consumes less fuel per mile driven than driving at faster speeds. Environmentally, this means that fewer air pollutants are generated for every mile driven. From a resource conservation perspective, less fuel consumed per mile driven obviously results in less fuel used in total. That was the driving force of the act after all.

You might wonder why driving at 55 mph consumes less fuel for every mile driven than driving at 65 mph, or 70 mph. Basically, there are four major factors that determine how much energy your car's engine requires to move you down the road. They are

1. The energy required to accelerate the car from a standstill to its cruising speed
2. The energy required to overcome the internal or rolling resistance of the car's drive train
3. The energy required to overcome the wind resistance the car encounters as it motors down the highway
4. The energy required for occupant comfort, including air conditioning, radio, and lights

Physics tells us that accelerating an object requires energy that is proportional to the mass of the object, the time duration of the acceleration, and the rate of acceleration squared. Thus, assuming a specified rate of acceleration, it takes a certain amount of energy to accelerate our car from being stopped to traveling at 55 mph. It takes a little more energy to accelerate to 70 mph as it takes a little longer to reach 70 mph as opposed to 55 mph. As we discussed in an example in Chapter 9, it also takes more energy if our car is heavier, 3000 lb versus 2500 lb, for example.

The energy required to overcome internal or rolling resistance has to do with the friction between the internal components of the engine and drive train. The engine's pistons rub against the cylinder walls of the engine block with every revolution the engine makes. Even though these surfaces are coated with oil, there is some friction resistance or friction energy that must be overcome in order for the piston to move.

Likewise, the engine's camshaft and crankshaft rotate in bearings and create friction. Notice that, even when our car is stopped at a traffic light or stop sign, your car's engine continues to idle. It consumes fuel just to keep the engine turning even though the car itself is stationary. Friction also exists in our car's transmission, and differential gears that connect the drive shaft to the drive wheels. It also exists in the wheels themselves as they contact the pavement. This is sometimes called rolling resistance. Granted, this rolling resistance, or the friction energy present in transmission, axles, and wheels only applies when your car is in motion, but it still consumes energy and requires fuel to keep the car in motion. If we turned off the engine while we were cruising down a flat stretch of highway, our car would eventually come to a stop, courtesy of rolling resistance.

Wind resistance is probably the main culprit that requires our car to use more energy to drive 70 mph versus 55 mph. The reason is that wind resistance is not merely proportional to the speed at which our vehicle travels but rather is proportional to our speed cubed. Here is the typical formula used to describe the power required to overcome wind resistance:

$$\text{Power} = 0.5 \times \text{air density} \times \text{speed}^3 \times \text{area} \times \text{coefficient of drag}$$

The air density refers to the weight of air per unit volume, pounds per cubic foot or kilograms per cubic meter, for example. It is sometimes hard to imagine air having mass as we move through it pretty effortlessly, but it does. Think of the air hitting the sail of a sailboat and pushing it along. We have also heard the term *thin air* referring to areas of high altitude where the density of the air is less. Area refers to the frontal area of our car in this example. A large frontal area, as in a large truck, for example, presents a large expanse for air to contact. A low, sleek sport coupe has a much smaller frontal area to intersect the wind. The coefficient of drag has to do with the shape and surface characteristics of our car. Well-designed cars, those with smooth edges and graceful curves, tend to have less wind resistance and generally can reach higher speeds. Makers of sports cars pay particular attention to the coefficient of drag of their designs.

So, if we increase our speed 15 mph from 55 mph to 70 mph, we have increased our speed by a factor of 1.273. This factor is simply the ratio of the speeds calculated as 70 divided by 55. The power required to overcome the additional wind resistance increases by a factor of 1.273 cubed, or 2.016. This essentially means that it takes twice as much power to overcome wind resistance at 70 mph than it does at 55 mph. That is the basic reason the speed reduction called for by the Emergency Highway Energy Conservation Act would be effective at conserving gasoline.

The energy required for occupant comfort is basically the same for both speeds. The radio and lights do not care how fast you are going. The air-conditioner might even work slightly better at 70 mph considering its condenser might be better cooled with increased airflow. Basically though, there is little to no impact for these auxiliary energy needs.

So the slower speed that we are forced to travel does pay off in both the environmental degradation and resource conservation pillars. Physics and the laws of nature are resolute in requiring us to expend more energy and create more pollution to move our 1970s era automobile down the road at a faster pace. Slowing down does have measurable benefits for both these pillars. I did not select a +2 score for either pillar because there were other options that we could have selected that would have saved even more fuel. Carpooling, mass transit, walking, biking, or measures to reduce the number of miles driven all could have potentially had a bigger impact on resource conservation, at least theoretically. Perhaps it is very unrealistic to assume that we could require vast numbers of motorists to carpool or take the bus as easily as we could require them to drive slower. That is a fair argument, but the idea of a rather modest increase in fuel economy rating a +2 on the resource conservation pillar seems like setting the bar too low.

The economic progress pillar gets a mixed score of 0/+1. In general, I would tend to go along with the +1 score due to the reduced cost of gasoline that goes along with the extra miles per gallon or kilometers per liter that we get from driving slower. However, as mentioned previously, it also takes longer to reach our destination if we drive 15 mph slower. A 500-mile trip to visit Grandma takes 7 hours and 9 minutes of drive time at 70 mph but 9 hours and 6 minutes at 55 mph. That is two extra hours to reduce your car's fuel consumption for the trip. I can see how some would view this as an economic penalty as well.

The personal benefit pillar receives a −2 score in my analysis. Most of us do not appreciate being told we have to slow down. We were doing just fine driving at the old speed limits, and 55 mph feels so slow when you are used to going 70. Many people bought cars specifically to go faster than 55 mph, and suddenly this became illegal. That was a drag, pardon the pun.

So, the benefits identified in the environmental degradation and resource conservation pillars are negated by the strongly negative personal benefit pillar, leaving the economic progress pillar score to stand as the total. This mixed 0/+1 score is appropriate, I think, for the situation in that some would begrudgingly accept the limitation on speed while others thought it an undue intrusion into their freedoms. It is interesting to note that this "national speed limit" has faded into history, and now states once again have the latitude to set higher speed limits.

This last example is indicative of how a neutral total score on a pillar analysis can signal an unsustainable action. The citizens of the United States ultimately disagreed with the provisions and restrictions of the Emergency Highway Energy Conservation Act of 1974. It points out that we probably will not agree with governmental decisions on really big environmental issues all the time. After all, many of us seem to take incredible joy in disagreeing with our government on most everything. We love to complain, and we have gotten pretty good at it. Unfortunately, the really big environmental issues will rarely be solved without some form of government action or intervention. As we discussed previously, the scope, complexity, and range of effects of the really big issues is usually great enough to require some type of concerted, even imposed, response. The pillars can help us figure out if there is good reason to disagree, and how important it is for us to make ourselves heard.

This is critically important because there are many groups and individuals weighing in on environmental issues who do not seem to care very much about the practicality of solutions. There are many environmental advocates who do not mention natural resources, or economics, or how some proposal to benefit the environment would hurt you or me. A practical environmentalist needs a gauge to quickly know if the latest gotta-do-it-now solution to the latest environmental calamity is really a good idea.

The pillars are the gauge, of course. I bet you saw that analogy coming. The pillars are the sanity check and a good prediction of effectiveness. Positive overall tallies on the pillars when applied to individuals mean that the benefits of the proposed action are reaching all the way down to each of us and favorably impacting our lives. Widespread negative pillar scores would inspire resistance and make it much less likely to achieve the desired environmental gain. Highly negative pillar scores should inspire resistance. Why should we go along with some idea when it represents a detriment to most of us? Why should we invest our energy and our wealth into an idea that brings widespread hardship or that is doomed to fail because of a lack of participation?

Strongly positive or negative overall scores from the pillars might encourage us to engage in the decision making around the issue. This is the hard part, of course. It is much easier to complain than to become involved. It is also difficult to get involved if we are not truly clear on where we stand regarding the issue to begin with. Hopefully, the pillars can alleviate that problem.

While this book is not intended to be a primer on interacting with our governments on environmental issues, getting our voices heard and respected is a natural next step. There are many ways to do this. In the United States, it is fairly easy to contact your congressional representative, and local politicians are usually very accessible. Many governmental programs and proposed legislation require public comment periods and invite our participation in this way. Newspapers regularly publish editorials, and Internet blogs are wide open. I will leave it to the reader to choose the best and most comfortable avenue for their situation.

Besides making up our own minds on environmental issues, the pillars can be useful to gauge whether a proposed action reasonably takes account of individual preferences and concerns. We are justified in asking if some "pillar-like" analysis was done by the action's proponent. For example, if a new federal pollution control law is proposed, has the federal government, the USEPA in this case, obviously considered and calculated the impact on an individual living in a Chicago loft apartment, or someone strolling the beach in Miami, or the recluse hiding away on a mountaintop in Montana? Will the proposed action be felt by citizens across the nation, or will it primarily affect a certain region of the country or a specific sector of the population? Could it potentially cause widespread hardship?

Very often, cost–benefit analyses are done to justify major environmental programs. This is different from the pillars though, in that cost–benefit analysis does not necessarily view things from an individual's perspective. It is more of a top-down approach, whereas practical environmentalism and the pillars definitely have a bottom-up perspective. The best environmental actions, the ones that make a truly positive impact, are founded on and grounded in benefits to individuals.

In the next chapter we will use the pillars to examine some big issues where significant benefit to individuals is hard to find. These are the sacrificial issues, and they deserve a very thoughtful analysis.

REFERENCES

Sierra Club. 2004. Hetch Hetchy, Time to Redeem a Historic Mistake. http://www.sierraclub.org/ca/hetchhetchy/ (accessed September 23, 2010).

State of California. 2006. Hetch Hetchy Restoration Study http://www.water.ca.gov/pubs/environment/hetch_hetchy_restoration_study/hetch_hetchy_restoration_study_report.pdf (accessed September 24, 2010).

National Science Foundation. 2010. America's Investment in the Future, Science on the Edge: Arctic and Antarctic Discoveries. http://www.nsf.gov/about/history/nsf0050/arctic/ozonehole.htm (accessed September 24, 2010).

Molina, M.J. and Rowland, F.S. 1974. Stratospheric sink chlorofluoromethanes: chlorine atom catalized destruction of ozone. *Nature*: 249, 810–812.

U.S. Department of Transportation (USDOT). 2010. Eisenhower Interstate Highway System—Frequently Asked Questions. http://www.fhwa.dot.gov/interstate/faq.htm#question13 (accessed September 25, 2010).

11 More Really Big Issues— The Sacrificial Ones

Sacrifice gets a bad rap. It sounds unpleasant, unwanted, and really unnecessary. It implies giving up something dear. Would not we rather focus on fun and enjoyment, profits and progress?

However, sacrifice has a positive side if we think of it in slightly different terms. I like to picture it in my mind as disciplined action and behavior in order to achieve a greater future good. While it might not be particularly pleasant in the present, it prepares us to prosper in the future.

We practice this type of sacrifice all the time. Athletes spend countless hours practicing and training. They run laps, lift weights, and perform drills over and over in an effort to improve their performance and prepare for success during competition. No doubt there is some enjoyment in these activities, but they are not painless, or easy. Sweat drips off their bodies, muscles strain, and their lungs struggle to inhale enough oxygen to meet the body's needs. Those athletes who are willing to sacrifice more of their time and energy to their cause tend to be more successful.

Good students make sacrifices as well. They devote hours and hours to reading and learning and practicing through homework. They are exercising their minds and stretching their knowledge. They give up television, computer games, and time with friends to focus on their studies. Unlike the athlete, their effort is not physical but mental. Yet it is generally not easy. It would be far easier to relax, have fun, and not think so much about improving their mind for the future.

In both of these simple examples we sacrifice current pleasure for future gain. It is an investment in ourselves and pretty easy to justify. Sometimes, though, we sacrifice for others. Parents do this regularly. They work long hours to provide for their family. They give up their own leisure time to invest in their children's activities. They constrain their expenditures to save money for their children to go to college.

We often make sacrifices for people we care about—our family members, our friends, and sometimes even for strangers in need. Sacrifice is a normal occurrence within our culture and one of humanity's defining traits. In this chapter we will explore the extrapolation of our common notion of sacrifice to environmental issues.

The sacrificial issues are the ones with negative scores on the pillars, a negative total score that is. It is quite normal to have some pillars score negatively and some positively within a pillar analysis of a proposed action. This is an indication of the trade-offs and compromises that are part of many of the decisions we make. When our analysis results in a negative overall score, going forward with the considered action represents a sacrifice.

Sacrifice within the pillars is a tough sell, and purposefully so. This relates back to Chapter 9 where we discussed the issue of sustainability. The point made then was that negative scores tend to be unsustainable and contrary to practical environmentalism, which seeks environmental benefit in a sustainable fashion. Practical environmentalism seeks to highlight negative scores as a recommendation to find an alternate course of action.

We often hear that our current lifestyle itself is unsustainable. We are told that the cars we drive pollute the atmosphere with smog and cause us to suffer poor health. We learn that the fossil fuels we burn to generate electricity or heat our homes cause environmental damage via global climate change. We fear that the genetic engineering that we perform on our crops and food supply risks the development of unknown powerful pests that could destroy our natural ecosystems. We are told that sacrifice is necessary for our survival. We are told that we must change our actions, our behaviors, and even our thoughts.

This call for sacrifice is common. With every new environmental calamity or dire scientific warning, the call becomes a little louder. Change your lifestyle. Give up those things that cause distress to the environment. Save the earth!

In these cases where we are specifically challenged to accept some level of sacrifice, we can expand our use of the pillars to evaluate the degree and distribution of such sacrifice. Where normally we would take the negative result and reject the proposed action as not being in our own best interest, we will now accept it, at least for a little while. We will give ourselves the opportunity to dig a little deeper and gauge whether the sacrifice is justified, and if so, it is fairly applied.

We could consider what the pillar scorecards of many individuals would look like. If we took a survey of a representative slice of our population, would the pillars predict a widespread negative reaction or merely a small, localized negative reaction? We could do this as a simple mental exercise, or we could actually conduct a survey and ask people to fill out the scorecard. This puts the issue in a clearer perspective and helps us understand if a request for sacrifice is warranted.

We ended the last chapter by emphasizing individual benefit as a cornerstone to effective environmental action and progress. Some may see this as a particularly selfish approach in evaluating environmental concerns. The really big issues will require some sacrifice after all, right? Americans should stop driving big gas-guzzling, air-polluting automobiles. We should allow our homes to be hotter in the summer and colder in the winter. Let us forget watering the lawn, and we really do not need all those chemicals the lawn care companies spray on our yards. Pay a little more for electricity, and it can be generated from renewable sources like windmills and solar panels. There is a long list of environmental improvements we could make if only we were willing to sacrifice a little bit.

While sacrifice may indeed be required in some situations, there should be a good reason for it. The Pillars allow us to include both in our evaluation and let us decide for ourselves if the sacrifice is a worthy one. We should pay close attention to which pillars produce negative scores. The general presumption is that there will be a positive score in the environmental degradation and/or resource conservation pillars, and negative scores in the others. This is what we would expect if we are called upon to make a sacrifice for the good of the environment. Negative scores in the human pillars of economic progress and personal benefit represent the expected monetary cost and change in lifestyle or standard of living that we would generally expect to

TABLE 11.1
Limits to Reasonable Sacrificial Action

Pillar	Score
Environmental degradation	+1/+2
Resource conservation	−2
Economic progress	−2
Personal benefit	−2
Total	−4/−5
Indication	OK, but only if it is really important

go along with sacrificial action. A negative score in the resource conservation pillar typically indicates a high cost in energy to accomplish the desired environmental benefit. If even the environmental degradation pillar reported a negative score, we should strongly question why we are considering the proposed action in the first place. After all, a positive impact to the environment is supposedly the driving force behind environmental activism. Environmental actions that result in increased environmental degradation seem in most cases to be obviously counterproductive.

Highly negative total scores from the pillars more likely represent ill-founded action rather than a call for sacrificial action on behalf of environmental benefit. In general, the most negative total score that could be considered reasonably sacrificial is in the −4 to −5 range. This is the case where we achieve an environmental benefit at a significant cost to the other pillars. The following scorecard shows this situation (Table 11.1).

The only way to get a more negative score would be for the environmental degradation pillar to score zero, −1 or −2. That would result in total scores in the −6 to −8 range. This range is beyond sacrifice, however, and probably indicates a bad idea from the beginning. Once again, why would we undertake sacrificial environmental action without a corresponding environmental benefit?

Now, maybe a few of us should sacrifice for the good of the whole. That is a noble human choice. As mentioned before, we often make sacrifices for our loved ones. Some even make the ultimate sacrifice of their own lives for the safety and security of friends, family, comrades, and country. Our armed forces and public services have many heroes who have made this choice. There is something truly selfless and noble in the character of these individuals, something in their souls that comes forward in the heat of the battle or the apex of the crisis and propels them to the greatest human service. This is both unusual and remarkable.

However, do environmental issues ever measure up to this standard? I dare say probably not. Even the really big environmental issues do not make it to this level. Lives are not in immediate danger, and there are probably many ways we could address the problem at hand. We have the luxury of time to make prudent choices. So, when we speak of sacrifice for the sake of the environment, we should be very mindful of the degree of sacrifice we ask of individuals and the number of individuals whom we ask to make this sacrifice. We should not expect general sacrifice without a corresponding general benefit. We should not expect a major sacrifice to be made by many for the sake of a few. To be effective and moral, sacrifice must be distributed fairly within our society and provide a meaningful benefit that is also distributed fairly within our society.

In the last chapter's examples of banning CFCs and implementing a cap-and-trade program to reduce sulfur dioxide emissions, the degree of individual sacrifice was relatively minor and the benefits were widespread. Both examples had negative scores in the economic progress pillar that were overshadowed by positive scores in environmental degradation and personal benefit pillars. These would not be considered as sacrificial actions but rather actions that had some sacrificial components within them. These actions asked us to pay a little more for the things we buy in order to have cleaner air and a healthier environment. However, the pillars showed us that indeed this was a good bargain to strike. It was worth it for us to pay for these improvements. Let us now consider a couple of cases where individual sacrifice is significant not just within a particular pillar but also in total.

For example, the United States Department of Energy (USDOE) has constructed a nuclear waste storage facility at Yucca Mountain, Nevada. Yucca Mountain is located in southern Nevada, approximately 80 miles (129 km) northwest of Las Vegas in a sparsely populated area on land owned by the federal government. The repository is constructed deep underground within the mountain itself. Radioactive waste from nuclear power plants across the United States would be sealed within underground chambers to protect humanity from ever coming in contact with the deadly radiation. Figure 11.1 represents the essence of this plan.

In 2002, the US Congress approved the Yucca Mountain site as the location for the country's nuclear waste (Gil and McKinnon 2002). Since then the USDOE has begun the process of obtaining permits and completing engineering work for construction to begin. The USDOE filed a license application with the US Nuclear Regulatory Commission (USNRC) in 2008 to operate the site as a nuclear waste repository (USNRC 2008). The plan is to transport nuclear waste from individual power plants scattered across the country to this central depository in southern Nevada for "permanent" storage and safekeeping. Here the waste would be stored for the thousands of years required for the radioactivity in the waste to decay to "safe" levels. Here all the country's nuclear waste can be kept under the watchful eyes of the federal government and protected from accidental or purposeful release. The project is currently subject to significant controversy, debate, and legal action, including reductions in funding for the project (USDOE 2010).

Let us apply the pillars to the situation and see what we come up with. The status quo for our analysis will be the current practice of storing spent radioactive nuclear fuel at many individual locations across the country. There are approximately 121 nuclear waste sites throughout the country (*Washington Post* 2009), most of them nuclear power plants (Table 11.2).

Environmental degradation gets a zero score in our analysis. The Yucca Mountain proposal essentially moves an environmental headache from one part of the country to another. Granted the "headache" would be deep underground in Nevada, but we have to get it there first. We have to pack it up and move it halfway across the country in some instances. There would be some small risk to the environment in shipping this waste. One significant spill could render an area uninhabitable. This risk must

FIGURE 11.1 Nuclear waste storage at Yucca Mountain. (Illustrated by Mark Benesh. With permission.)

TABLE 11.2
Yucca Mountain Nuclear Storage Site

Pillar	Score
Environmental degradation	0
Resource conservation	−1
Economic progress	−2
Personal benefit	−1
Total	−4
Indication	Yucca Yuck!

be balanced against the environmental risk in leaving it alone right where it currently sits. This seems more or less a break-even proposition to me. If you happen to be a nuclear physicist, maybe you could perform a detailed risk analysis and come up with a more accurate answer. However, in most cases, the pillars do not require highly accurate answers, and as you will see in a moment, that is the case here as well.

Resource conservation is scored a −1. This may be overreaching a bit, but the project will require a good bit of cross-country transportation. A significant amount of energy has already been expended to dig deep inside the earth to develop the underground storage chambers within Yucca Mountain. This is a sunk cost, however, as the energy to construct the repository has already been spent. There will undoubtedly be an ongoing energy requirement to operate Yucca Mountain though. All in all, the energy required to transport and store nuclear waste is probably not a huge amount of resource consumption when compared to all the gas we burn driving our cars or all the fossil fuels we consume to heat, cool, and power our homes. I certainly would not quibble if you wanted to give this pillar a −1/0 score. It does seem decidedly negative though, so I do not see a zero score here.

Economic progress gets a −2 score. The government has expended huge sums of money to develop this site. More money will be spent on shipping the nuclear waste in specially designed crash-proof containers. Money will be spent on lawsuits that will inevitably arise when people learn that the nuclear trash highway goes right through their town. A project of this scope and size, and one that is highly contentious, will not come cheap. While it will create a few jobs, it does little to improve economic efficiencies or improve our nation's productivity. This one seems like a big economic loser.

Personal benefit gets a −1 score in our analysis. Why should we move nuclear waste around the country if we do not have to? Nuclear waste would be rolling down the highway right next to Grandma as she drives her Plymouth to visit the grandkids. It would be riding the rails through the small towns and farm fields of our nation's heartland. We do not think many people will be real comfortable with this idea. Any personal benefit comes from a complex and extended string of logic saying that we are somehow safer by concentrating all the waste at one site rather than leaving it at its many current storage sites, typically at the nuclear power plants themselves. We are told that the many current storage sites were not designed to hold this waste for the many years needed for it to defuse itself. We are told that it would be easier to protect the waste from accident, sabotage, and terrorism if it were all buried at Yucca Mountain. Even accepting this safety premise, it does not seem to outweigh the transportation concerns. Again, I would not quibble if you wanted to score it −1/0.

So, our total tally with the pillars is −4, perhaps a −3 if you used both −1/0 modifications. We came up with no positive scores for any of the pillars. The alarm bells should be going off and the red lights flashing. The Yucca Mountain project certainly does not pass muster with the pillars. Not only do the pillars show this proposal to be sacrificial but they also indicate that this sacrifice comes about with little perceived environmental benefit. If there truly is some reason that warrants this action, the proponents of the action need to do a much better job convincing us that it is worth the level of sacrifice required from us.

Also in this case, the degree of sacrifice is not evenly distributed among Americans. People who live near nuclear power plants or on waste transportation routes would be most impacted. The other important question to ask is, "Who would benefit?" That's a tough question because there are no direct benefits, only a perception that somehow we would be "safer" with Yucca Mountain in operation. This perception is time dependent as well. I know of no places where nuclear waste storage poses an imminent threat to those living nearby. That may not be true in 50 years, but, are we willing to accept hardship and risk today in trade for an uncertain benefit many decades from now? What kind of answer do you suppose we would get if we surveyed Americans on this issue?

I am not really suggesting that we canvass the country and try to convince America that Yucca Mountain is a fatally flawed idea. That is a step beyond practical environmentalism. Advocating a cause can certainly be a noble endeavor and could easily follow from the appropriate application of the pillars. Practical environmentalism is content, however, with encouraging people to put the proposed environmental action under reasonable and balanced review with regard to each of the four pillars. There remains a place for soapboxes and sit-ins, for marches and mass advertising, and for voting your conscience as well as your pocketbook. Practical environmentalism is meant to inform these next steps and provide a firm foundation for sustainable progress.

It is relatively easy to get carried away with these types of sacrificial issues. As the negative scores mount with each pillar we examine, we may become a little upset. The pillars paint a very negative picture of nuclear waste storage. Everyone already knows that nuclear waste is a very bad thing, right? Remember, however, that the pillars are meant to address very specific questions. In this case, that question was where we should store the radioactive waste material. The question was not whether nuclear waste material is good or bad or even tolerable. Unfortunately, the sad truth is that nuclear waste material is already with us and will likely be with us for thousands of years. Unless someone can unlock the secret to diffusing the deadly radiation or figure out how to safely jettison it into outer space, or even send it to be consumed within the sun itself, our grandchildren's grandchildren will likely be struggling with this issue in ways that may not be too different from our current struggles.

Our question was, Does Yucca Mountain make sense from a practical environmentalism perspective? Does Yucca Mountain reduce environmental degradation at a cost in resources, economics, and personal benefit that we are willing to bear? Our answer to this question was unequivocally NO. Proponents of the Yucca Mountain site should take notice of the sacrifice they request from us. They should infer that their proposition may not be sustainable. They should consider modifying their proposal to either lessen the negative impacts of each pillar or, at least, distribute the sacrificial impacts in a manner that is equitable and that can be seen as necessary by the citizens affected.

Let us look at another example. This one is a little less well known but still very interesting. It has to do with water.

In the western Great Plains of the United States there exists a huge underground aquifer known as the Ogallala aquifer. This is basically a vast underground lake that extends from South Dakota and Wyoming as far south as West Texas. Much of the drinking water for people who live in this part of the country comes from the Ogallala. It is also the source for most of the water used to irrigate cropland in the region. Irrigation is considered a necessity for many crops because of generally scant precipitation in the area. The water table of the Ogallala has been decreasing for years, and some fear it could dry up within the next two or three decades (Little 2009).

TABLE 11.3

Ban Irrigation from the Ogallala

Pillar	Score
Environmental degradation	0
Resource conservation	+1
Economic progress	−2
Personal benefit	−1
Total	−2
Indication	Do we have to?

Let us assume that the United States Environmental Protection Agency (USEPA) proposed a law banning crop irrigation using water from the aquifer. Let us apply the pillars and see if this is a sacrificial proposal (Table 11.3).

Environmental degradation gets a zero score in our analysis. Does that surprise you? We are draining an aquifer after all, and that cannot be good. Indeed it is not good, but the detriment is to us humans instead of the environment. In previous examples of ozone holes, air pollution, and nuclear waste, the potential for environmental damage extended beyond human impacts to plants and animals. In all these cases, there was some significant potential damage at the earth's surface, which can be a pretty crowded place, ecologically speaking. In the case of the Ogallala, however, the impact is subsurface and deep enough where there is not much biological activity. It is hard to see that the water table level in the Ogallala has much of an impact on the environment. Any impacts are likely to be small and so a zero score seems warranted.

Resource conservation is scored a +1. Water is definitely an important resource, and maintaining a suitable amount of the resource is by definition a plus. I did not give it a +2 score because, once again, we are dealing with some of the confounding factors mentioned in Chapter 1. While there is hard data showing water table decline, the action of banning irrigation comes from a fear based on a prediction. Whenever we are dealing with predictions of future states, I get very cautious with my scores and generally go with a 1 over a 2.

Economic progress gets a −2 score. Banning irrigation would likely force many farmers to plant less productive and less profitable crops. Some would probably go out of business. There would undoubtedly be an economic ripple effect as much of the local economies in the region are heavily based on the agriculture industry. As profits dwindle in the farming business, all sorts of businesses feel the pinch as well. Bankers, merchants, grocers, suppliers, and many others may see their incomes drop. People start moving away in search of profitable employment, and real-estate markets suffer. The tax base begins to erode, and public services start to suffer, too. This one seems like another big economic loser.

Personal benefit gets a −1 score in our analysis. This score is meant to reflect the range of impacts across the spectrum of individuals within the region. If you are an affected farmer, a −1 score is easy to justify and a −2 is probably more appropriate.

If you are not directly involved in agriculture, the negativity might not be so strong and you might choose −1 or even zero. It is good to have water, of course, but our fear of running out of water seems like a long way off. Reduced incomes and poverty would be a more immediate concern. The negative economic effects would impact our personal well-being more than the fear of a future water shortage. Some environmentalists may even choose to score this pillar positively. A score of −1 seems like a good general compromise.

Our total tally with the pillars for banning irrigation turns out to be −2. It is definitely a call for sacrifice in this case in the name of resource conservation. But is it warranted? Again, let us look at who pays the price and who reaps the benefits.

The Ogallala aquifer issue is ultimately a regional issue. It primarily impacts the local population, so it is the local people who pay the price and reap the benefits. The interesting part of the issue is that those who pay the price and those who reap the benefits are identical geographically, but they are separated in time. The price of economic disruption would be immediate, while the benefits of a continuing readily available water supply accrue to children and grandchildren one or two generations distant. Those who pay the price are only indirectly linked to those who reap the benefits through family lineage.

As mentioned previously in this chapter, the main benefit of the pillars with this type of issue is to identify it as sacrificial and prompt us to search for alternatives. Let us do just that. Let us examine another proposed action with the intended benefit of stabilizing water availability in the Ogallala.

Instead of banning irrigation altogether, let us propose that we will manage the rate of water withdrawals to stabilize the level and quantity of water in the aquifer. Furthermore, our proposal will follow the cap-and-trade protocol where existing water users are granted water withdrawal rights in proportion to their current usage rates. These water withdrawals will be less than current usage, but this decline will be fairly distributed and all will be guaranteed access to some amount of water. A central authority, perhaps a USEPA regional office, will be charged with administration of the program.

This cap-and-trade program for Ogallala water will apply to all major users of water. Municipal water systems and industrial users will be subject to these new rules as well. In addition, exercise of these water withdrawals will be associated with a reasonable fee paid to the regional USEPA office charged with the administration of the program. A part of these fees will be allocated to the USEAP for reasonable administrative expenses, but the majority will be paid into a fund with the express goal of encouraging water conservation efforts within the region. This fund could be designed to pay out rebates and incentives to individuals, businesses, and even municipalities that reduce their water use by installing more efficient equipment or systems. Homeowners may get a rebate for installing low-flow showerheads in their bathrooms. Farmers could get a rebate for using highly efficient drip irrigation systems in their fields instead of the massive spray irrigation equipment that loses a lot of water through evaporation. Manufactures that optimize their production lines to use less water could get a rebate check. Even municipal water systems could get a piece of the pie. Perhaps they could get a grant from the fund to cover the costs of a program to promote water conservation methods for their customers or offer free water conservation audits and suggestions.

The fees associated with withdrawing water from the Ogallala aquifer would probably only be assessed to large consumers of water such as municipal water systems, major industrial users, and large farms. These fees could be designed on a sliding scale with the unit cost of water increasing as the usage rate increases. This could create a powerful incentive, in and of itself, to use less water. This tiered structure of water rates might charge these large customers $1000 for the first million gallons used, but $1500 for the second million gallons. Perhaps the third million gallons used in any month might cost $2000. As more water is pumped from the Ogallala, the average cost of this water increases and creates an incentive that could justify water conservation efforts.

Of course, our proposed program would limit the overall withdrawal of water from the aquifer. Water meters with remote sensing capability could be required of all large wells that draw from the Ogallala. The owners of these wells would have a certain number of allowances they could use to withdraw water. If well owners decided to implement conservation measures and not require all their allowances, they could sell them to others who needed more water than their own allowances would permit. This is the "trade" part of the cap-and-trade strategy and is meant to promote economic efficiency by allowing water use where it has the greatest value. The overall withdrawal limit from the Ogallala could be managed as well with the maximum water withdrawal potentially changing year by year. This allows fine-tuning of the hydrological cycle so that water withdrawals can be matched to the aquifer's natural water recharge rate, which may vary over time. The end goal of the program is to stabilize the aquifer while still meeting the water needs of the region.

Let us see how our proposed cap-and-trade approach to manage Ogallala water resources fares under scrutiny by the pillars (Table 11.4).

Environmental degradation still gets a zero score in our analysis. The logic from the previous example holds here as well. It remains hard to see that the water table level in the Ogallala has much of an impact on the environment.

Resource conservation gets a modified +1/+2 score in this example. The cap-and-trade program, with its water monitoring abilities and its flexibility in setting annual discharge limits, is seen as an effective management strategy in maintaining a sustainable water supply in the region. The slightly more positive score here reflects the

TABLE 11.4

Cap-and-Trade Water Rights to the Ogallala

Pillar	Score
Environmental degradation	0
Resource conservation	+1/+2
Economic progress	−1
Personal benefit	0
Total	0/+1
Indication	That is better

cap-and-trade program's design of using water access fees to fund improvements in water supply and conservation efforts in general. These efforts could reach beyond the agricultural water users and deliver improvement to municipal and industrial users as well.

Economic progress gets a −1 score, a step better than the −2 score received by the banning irrigation proposal. Many farmers would certainly incur increased costs, but they would have access to water. Hopefully, these costs would be more fairly distributed across the economic spectrum and would deliver a benefit in ensuring a dependable, even though reduced, water supply for their crops. Some less productive and less profitable crops would probably be pulled from production, but this would occur in a much more gradual fashion than if irrigation were banned altogether. The negative economic ripple effect would hopefully be much less severe as water restrictions are phased in over time. Perhaps other parts of the economy may even flourish once the uncertainty of future water supplies is alleviated. Improvements in water use efficiency, funded by the cap-and-trade program, may provide economic benefit to some through savings associated with reduced water usage. With this scenario of reduced economic penalty, some potential of economic gain, and more certainty in general, I think a −1 score is reasonable.

Personal benefit gets a zero score in our analysis, again reflecting a range of impacts. If you are an affected farmer, a −1 score is still easy to justify, but this cap-and-trade approach is certainly less onerous than an outright banning of irrigation water. Residential users may benefit from the incentives created by the program to conserve water. It is also generally good to have a greater level of certainty around a secure long-term source of water. Also, if managed wisely, the program would reduce the negative economic ripple effects that could happen with an irrigation ban.

Our total tally with the pillars for managing water flow from the Ogallala is now zero to slightly positive and represents a significant improvement over the −2 score for banning irrigation water to be drawn from the aquifer. What was seen as an obviously sacrificial attempt to minimize resource depletion has now been modified and could be perceived as a reasonable and sustainable program to manage long-term water supplies for those who depend upon the Ogallala reservoir of underground water.

This case study of the Ogallala demonstrates the value of the pillars. This type of analysis lets us move from sacrificial efforts to sustainable progress with relative ease. All that is required is our own creativity in identifying feasible alternatives to the status quo. The status quo of the Ogallala is seen as dire by many. In this instance, it is easy to jump to the conclusion that sacrifice is indeed necessary. In general, the more dire the status quo, the easier it is to justify sacrificial action. Practical environmentalism, however, always challenges this type of knee-jerk reaction to environmental calamity, for these calamities thrive within the confounding factors of Chapter 4. Confusion is never an ally of environmental progress.

The pillars allow us to minimize the confusion created by perceived environmental calamity. We can quickly run a number of alternatives through the pillars and get a basic understanding of which options are sustainable and which are sacrificial. There are, of course, other options to the Ogallala crisis than the two discussed here. These other options could be analyzed as well, and we would have a better representation of the actions we could take to improve this situation.

The pillars show us that there is truly little need to quickly and emotionally react to environmental crises. We really do have the luxury of a reasoned response. The next chapter delves into the biggest environmental crisis of our time—global climate change. Let us apply practical environmentalism to this perceived crisis and begin to understand how different remedies compare to one another through pillar analysis.

REFERENCES

Gil, A.V. and McKinnon, B.L. 2002. Regulatory Framework for the Geologic Repository at Yucca Mountain Nevada, U.S.A. ANS Topic Meeting on Spent Nuclear Fuel and Fissile Materials Management, Draft Paper #32. http://www.osti.gov/bridge/purl.cover. jsp;jsessionid=6F95A48B0A239541BE0366AE97B42B94?purl=/802602-rPfEP8/ webviewable/ (accessed September 26, 2010).

US Nuclear Regulatory Commission (USNRC). 2008. DOE's License Application for a High-Level Waste Geologic Repository at Yucca Mountain. http://www.nrc.gov/waste/hlw-disposal/yucca-lic-app.html (accessed September 26, 2010)

US Department of Energy (USDOE). 2010. Office of Civilian Radioactive Waste Management. http://www.energy.gov/environment/ocrwm.htm (accessed September 26, 2010).

Washington Post. 2009. If Nuclear Waste Won't Go to Nevada, Where Then? http://www .washingtonpost.com/wp-dyn/content/article/2009/03/07/AR2009030701666.html (accessed September 26, 2010).

Little, J.B. 2009. The Ogallala aquifer: Saving a vital U.S. water source. *Scientific American* (Special Edition, March 30, 2009).

12 The Pillars and Global Warming

Global warming is such a high-profile issue that it deserves its own chapter. We will briefly look at its history and theory and then apply the pillars to hopefully make some sense of several of our potential responses.

Global warming, or climate change as it is sometimes called, is the main environmental issue of our time. The theory goes something like this. We humans are burning too much fossil fuel and liberating vast amounts of carbon dioxide (CO_2) into the atmosphere. The CO_2 acts like a one-way heat filter in that it allows sunlight to pass freely from outer space into our atmosphere but prevents heat from the planet from reradiating freely into space. Other atmospheric gases, some natural and some man-made, also exhibit this property. Scientists tell us that, without the heat-filtering processes of these gases, the earth's average temperature would be approximately 30° Fahrenheit (15° Celsius) cooler (Le Treut et al. 2007). Brrr …

The term "global warming" can be traced back to at least the mid-1970s. Wallace Broecker wrote an article entitled "Climate Change: Are We on the Brink of a Pronounced Global Warming?" that appeared in the journal *Science* in 1975 (Broecker 1975). In the article he presents the argument that decades of global cooling will be replaced with decades of global warming due to increasing carbon dioxide concentrations. While Broecker was certainly not the first to study earth's climate, he did recognize the linkage between anthropogenic carbon dioxide emissions and global surface temperatures back in the early 1970s and predicted increasing temperatures. We have come to understand this linkage as global warming. Over the decades since 1975, global warming has increased in importance within our scientific community, our global politics, and our societies in general. In 1989 the Intergovernmental Panel on Climate Change (IPCC) was created by the World Meteorological Organization and the United Nations Environment Programme (IPCC 2010) to report on the status of the world's climate. Since then the IPCC has produced four major reports related to climate change, and continues to be a focal point for discussion, debate, and action around this global issue.

Global warming has created its own dictionary. We now have greenhouse gases, carbon footprints, and climate change. The term "greenhouse gas" is a reference to CO_2 and all the other gases that act much like the glass of a greenhouse. Water vapor (H_2O), methane (CH_4), nitrous oxide (N_2O), ozone (O_3), and chlorofluorocarbons (CFC) are other examples of these gases. Just like the glass walls and roof of a greenhouse, these gases allow sunlight to pass through but tend to block heat from escaping. This curious property of these gases is due to their atomic structure and the different intensity levels of the radiated energy. They tend to block radiated energy in the infrared range while they allow visible light to pass. Energy in sunlight is very intense as it is generated at the sun's surface at approximately 10,000°F (5538°C or 5811 K). Heat radiating from earth back into space is much less intense as it is generated at the earth's surface at a maximum temperature of approximately 100°F (38°C or 311K). The amount of energy transferred by radiation is a function of the temperature of the source raised to the fourth power. Comparing the absolute temperature in degrees Kelvin (K), the sun's temperature is 18.7 times hotter than the earth's temperature (5811/311 = 18.7), but the energy per unit area, or intensity, is 18.7^4 or more than 100,000 times greater.

Carbon footprint is another modern addition to our environmental dictionary. The term does not mean a black muddy mark on the ground, but instead, it is a reference to CO_2. Carbon refers to the carbon atom, C, in the CO_2 molecule. The idea behind the term is that the carbon footprint quantifies all the CO_2 released by a certain activity. We can calculate our own individual carbon footprint by adding up all the CO_2 released through our daily activities of living. We can figure out how much CO_2 is released by driving our car as well as how much was released in making the car. We can estimate how much is released by heating our homes, watching TV, going to school, making our clothing, or producing our food. We tend to focus on CO_2 because it exists in larger quantities than the other greenhouse gases and because it readily fits the global warming theory of human-induced climate change via fossil fuel use. Thus, our carbon footprint is a measure of how much we are contributing to global warming. From this standpoint, large footprints are bad, and petite ones are good.

Climate change is a pseudonym for global warming but also a term that can include other weather phenomena such as precipitation patterns and violent storm events. As such, climate change is a much broader term than global warming. It is not restricted to merely an increase in temperature. Climate change can include increased frequency and severity of droughts, floods, hurricanes, and blizzards. It can even include cooling temperatures in certain parts of the world. Climate change is a very inclusive and convenient term that allows humanity to accept responsibility for all sorts of weather events. This broader definition allows us to increase the size and scope of the issue and the upcoming catastrophe if we wish. It is also a way to hedge our bet and grant us an escape clause just in case the predicted temperature changes do not materialize as we expect. Taken to extremes, the inclusiveness of the term climate change basically allows any weather outcome to be promoted as resulting from humanity's impact upon the global climate.

There certainly is physical evidence to support the global warming theory. Atmospheric observatories around the world are tracking the concentration of CO_2 in the atmosphere and report an unmistakable increasing trend of CO_2 concentration. Figure 12.1 illustrates this point with CO_2 concentration data from Mauna Loa observatory in the Hawaiian Islands.

This chart shows a clear trend of increasing CO_2 concentration from 330 parts per million by volume (ppm_v) in the mid 1970s to nearly 390 ppm_v in the 2008–2009 time frame, an approximate 18% increase in 35 years. The chart also displays an unusual annual undulation, a "sawtooth" pattern, throughout that is typical of CO_2 concentration trends. The sawtooth pattern occurs due to the annual growth cycle of plants that depend upon photosynthesis. The process of photosynthesis uses

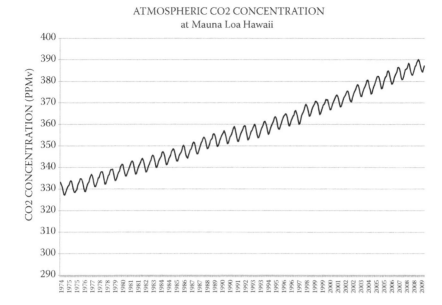

FIGURE 12.1 Atmospheric carbon dioxide concentration at Mauna Loa Observatory. (From Tans, P. 2010. Mauna Loa Monthly CO_2 Concentration Dataset. National Oceanic and Atmospheric Administration Earth System Research Laboratory. http://www.esrl.noaa.gov/gmd/ccgg/trends/ [accessed September 26, 2010].)

atmospheric CO_2 and the energy in sunlight to produce plant biomass. As the Iowa corn crop grows taller, some CO_2 is removed from the atmosphere and becomes part of the plant. As trees, grass, and other plants grow, the same thing happens. Taken together, plants throughout the Northern Hemisphere remove enough CO_2 to be measurable and impact the overall carbon dioxide trend in the annual cyclic pattern show in Figure 12.1.

The other critical piece of supporting evidence is the temperature record. Figure 12.2 shows global average surface temperature anomalies from 1880 through 2006. A temperature anomaly is the difference between a measured temperature and some average value. In the case of Figure 12.2, the average value is the temperature mean between 1951 and 1980.

I have superimposed two arrows on the data to indicate periods of generally increasing temperatures. These periods are from 1908 to 1943 and from 1964 to 2006. The more recent period of increasing temperatures aligns well with the CO_2 concentration trend of Figure 12.1. The time period between these two periods of temperature increase seems to show a slight cooling trend, which is in agreement with Broecker's analysis (Broecker 1975). Overall, the chart shows an increase of 1°C over the 125-year period shown. The trend is not a smooth increase, however, as we might expect from the global warming theory if atmospheric CO_2 concentration was the only critical factor influencing global temperatures.

In addition to the CO_2 and temperature data, we hear that glaciers are melting and retreating (IPCC 2001). There are reports of longer growing seasons and changes in

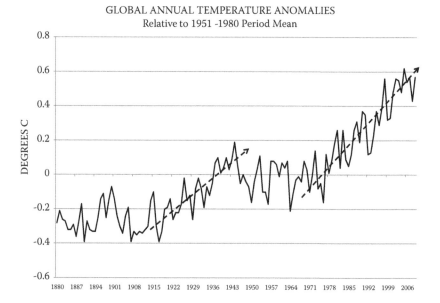

GLOBAL ANNUAL TEMPERATURE ANOMALIES
Relative to 1951 -1980 Period Mean

FIGURE 12.2 Global annual temperature anomalies. (From Hansen, J.E. et al., 2010. NASA GISS Surface Temperature [GISTEMP] analysis. In *Trends: A Compendium of Data on Global Change*. Carbon Dioxide Information Analysis Center, Oak Ridge National Laboratory, U.S. Department of Energy, Oak Ridge, TN. doi: 10.3334/CDIAC/cli.001.)

the types and distribution of species in some parts of the globe (Davis 2010). The weathermen tell us that the majority of the warmest years on record have occurred within the past 10–20 years, which goes hand in hand with the temperature anomaly data presented earlier in Figure 12.2. Oceanographers tell us that the oceans are warmer than before. Reportedly, the oceans of the world have recently been measured as 1°F (0.56°C) warmer than normal, with normal being the 20th-century average (NOAA 2010). As Figure 12.3 somewhat comically implies, the world seems to be getting hotter.

To be honest though, I have not noticed these changes. I have not felt them. Perhaps it is because I like it warm. I do know that, where I live in eastern Iowa, we just set an all-time record low temperature for Cedar Rapids, Iowa. Granted, an all-time record in this instance really means for about 150 years or so, which is as long as they have been recording temperature data in Cedar Rapids, Iowa. Still it is a curious twist of the data that we experience the coldest day in recent history during a time when we expect our environment to be getting warmer and warmer.

This unexpected weather record points out one of the difficulties in examining the global warming/climate change theory. We often confuse weather with climate. Weather occurs locally on an hour-by-hour basis. We check the weather to figure out how to dress our children before we send them off to school, or to find out whether we can squeeze in a game of tennis after work. Weather is local, variable, and time specific. Climate, on the other hand, is all about long-term averages. We might compare the amount of rainfall in our town during 2009 with what fell in 2008. We

FIGURE 12.3 An overheating earth. (Illustrated by Mark Benesh. With permission.)

might compare the cooling degree days (often a measurement for air-conditioning calculations) from one year to the next, or even one decade to the next. One particularly rainy or hot day does not necessarily mean that the whole year will be wetter or hotter than "normal." Climate change is much more concerned with what happens over a year's time than how cold it got in eastern Iowa one winter night.

However, we humans are very concerned about the weather. In the United States, a good portion of our local news programming is all about the weather. We can check the weather forecast on the Internet or have weather alerts sent to our cell phones. Our interest in the weather naturally extends to a curiosity about how global warming will affect our weather. It becomes very easy to confuse one with the other. It becomes deceptively easy to blame our local weather extremes on global climate change phenomena.

Weather is right now; climate is the past, present, and future. Climate change theory is very cognizant of this. Climatologists have a variety of data records and experimental methods to gain a better understanding of how our climate has changed and will change in the future. We can peer back in time to analyze past climates by examining historic temperature records, chemical composition of gases trapped in ice cores, and even tree-ring patterns. We can look ahead to the future with many complex computer models that attempt to predict what our climate will be 20, 50, or 100 years from now.

Our past can be very important in putting things in perspective. There is an old saying that goes something like, "Those who do not understand history are condemned to repeat it." There is a link to the past in the curious name of a town near where I live. The town is Coralville, Iowa. Coralville is a very pleasant community that borders Iowa City, Iowa, home of the Hawkeyes. Go Hawks! Interstate 80 runs right through Coralville on its trek from coast to coast. This is Midwest farm country, the land of corn, soybeans, and pigs. You might be wondering what is so special about this place to warrant a mention in this chapter on global warming. That is a good question. The town is special because of its name, Coralville. Apparently it got its name from fossilized coral found in limestone along the Iowa River. You might wonder how fossilized coral came to be in Iowa. Coral, after all, is a marine organism that is typically found in shallow tropical seas. We do not have many tropical seas in Iowa—none actually. The likely answer to this puzzle is that millions of years ago this particular part of Iowa was in fact the floor of a shallow tropical sea. The land that is now Coralville was under a bunch of warm water. Bright sunlight filtered down through the water, and coral thrived.

The very name of this eastern Iowa town suggests that our climate has changed before, long before we were around. We have also heard of prehistoric ice ages and how they influenced and formed our midwestern landscape in many places. So, living here in Iowa, I can imagine both extremes. If I could transport myself back in time far enough, I could find myself peacefully floating in warm tropical waters or alternatively shivering on some icy glacier.

This logic begs the question of whether climate changes because of natural events, human impacts, or both. Also, we must somehow factor into this logic the timescale that is appropriate for these events. None of us remember the last ice age or the Devonian period in prehistory when the corals of present-day Coralville were alive. You could argue that these pesky theories are simply irrelevant to our current discussion on global climate change. They occurred far too long ago to be considered as part of our current predicament.

Yet these historic climates that are drastically different from our present climate offer us a perspective on the scope and power of humanity's impact on the environment. It seems obvious that human actions had nothing to do with corals thriving underwater in present-day eastern Iowa or alternately massive sheets of glacial ice covering part of the Iowa landscape. It also seems likely that we humans are changing the earth's atmosphere by slowly but surely adding to the concentration of CO_2 that exists in the atmosphere.

There are two key questions that follow from these conclusions. First, what is the extent of natural climatic changes? Second, how large are human-induced climatic changes compared to natural ones? These are both very difficult questions, primarily because we simply do not understand why natural climate changes take place and how quickly natural changes can influence climate. We do not know why eastern Iowa enjoyed a tropical climate at one period in history, nor do we know why

Iowa suffered through an ice age during another. I pose these questions without the expectation of an answer, and I do not think that searching for an answer within the confines of this book would be particularly rewarding. These questions are best left to paleoclimatologists. Let them debate and develop the theories of ancient climatic events within the comfortable confines of academia.

The two preceding questions are interesting to me because they are essentially ignored in our current discourse on global warming. Perhaps they are ignored because they are so difficult. Why waste time on a question we cannot answer. Instead it seems that we have leapt to the conclusion that global warming is certain and real and we understand its causes and dynamics well enough to predict the future. How audacious we have become!

The great value in these unanswerable questions is to put this huge issue in a better perspective. It serves to contrast the reality of our knowledge about our climate with the certainty in which global warming is held in our modern society. There is so much we do not know, so much we do not understand, yet we pretend that we do.

A practical environmentalist should not hide from the issue simply because of uncertainty. Global warming is an environmental issue regardless of the fact or fiction of the theory itself. It is an issue simply because our society has accepted it, seemingly without question. Global warming, and humanity's responsibility for it, has become fully ingrained in our society and our psyche. It is commonly accepted as fact.

Global warming, however, suffers from nearly all the confounding factors mentioned in Chapter 4. It is a looming crisis founded on the fallacy of prediction and the assumption of future states. Its altered future, one of flooded cities, vicious storms, and mass starvation, feeds our paranoia. The uncertainty around this altered future as well as the range of measures of success, are huge, yet it receives far more attention than other environmental issues competing for our support. We are inundated with expert scientific opinions reporting the latest results from the computer models that divine our future. We hear the exhortations to curb greenhouse gas emissions to 1990 levels by the year 2030, or 2040, or 2050. We hear the slogans promoting the reduction goals—20% by 2020 or 50% by 2050. The propaganda is excessive, reasoned discourse scarce.

Human culpability may or may not be completely true. Global warming, or climate change, is still predominately a theory. Many consider the historical evidence of increasing atmospheric CO_2 concentration and increasing average temperatures to be strong and convincing, but this evidence itself is not proof of the theory. This evidence supports the theory but does not necessarily rule out other potential theories that may also account for an increase in temperature. The ultimate proof of the theory might be hard to come by. Normally we prove theories by being able to manipulate the critical input factors to the theory and obtain the expected outcome. That is to say that we prove the theory by experimentation. With global climate change, however, we are not in control of this grand experiment. We are merely along for the ride. We try to piece together data from the historical record as inputs and outputs of the theory, but these linkages fall short of absolute proof. The best we can probably hope for is evidence that either strongly supports the theory or exposes a great fault in the theory that would lead us to look for alternate explanations of the phenomena that we experience. Even this could be difficult, however, in that we do not experience our lives as a laboratory experiment. If the ice melts, sea level rises, and proliferation of violent mega storms does not materialize, perhaps it was because we heeded the warnings of global climate change and altered our behavior just enough to prevent the impending disasters. Or, this future state could also be explained by saying that our global warming theory was not quite right to begin with. We might have missed a critical factor or misjudged some natural dynamic that would also explain the unexpected result. It would be very difficult to know which of these two explanations was truer.

It sure would be nice to wait and see if the global warming theory was true or not. If we could wait 50 or 100 years, we could see if the earth's temperature really did rise as the computer predictions indicated it would. Of course, the problem with this approach is that if human-induced global warming is true, then the wait-and-see strategy brings upon us the consequences of the theory. The predicted damage would be done. By then, the Antarctic ice would have melted, sea levels would be higher, and current coastal areas might be well under water.

Alternatively, we could proceed under the assumption that global warming is real and begin to take action now. The risk with this approach is that if we are wrong, we will waste a lot of time, energy, and money. These resources could have been better spent in other ways to improve the environment or improve our lives. We could take the huge amounts of money that are pouring into global climate change research and advocacy efforts and redirect them to obvious and immediate needs. We could invest in clean and sanitary water systems around the globe that could save thousands, even millions, of lives over the next several decades. We could invest in the education of many young people who cannot afford to attend college. We could make sure that no child wants for food. There is no shortage of pressing needs and worthy causes. This essentially means that there is an opportunity cost in fixating our attention and energies on global climate change. Worrying about global climate change for the future does prevent some environmental benefit from being realized today.

Also consider that there are several ways we could be wrong under the global warming is real assumption. First, we could find out that CO_2 emissions really have nothing to do with atmospheric temperature. Second, we could find out that the theory is partly true but the expected effects are much less severe than we thought. Third, we could discover that there are natural forces at work causing the temperature to rise, and all our efforts to reduce CO_2 emissions are insignificant compared to Mother Nature's whims.

All of these errors are based around the idea that we simply do not understand how the earth's climate works and what makes it change. There is some truth in this assertion as there are many things we do not know about why the weather is what it is. Even in the short term we have trouble predicting the weather for more than a few days into the future. In the global warming discourse, we depend upon computer models of varying degrees of complexity and completeness to peer decades into the future to discern what our world may look like. How many times has humanity predicted the future with much success?

Yet, we do know many things about the climate, and we are very used to proposing theories and using crude models to gain some insight into how nature works. Indeed, it is the very nature of human beings to try to understand the workings of our world and how to leverage this understanding for our benefit. Therefore, discounting global warming merely because there is a great level of uncertainty is simply unsatisfactory. The high degree of uncertainty, however, should give us reason to pause and weigh our options very carefully. This is precisely where practical environmentalism can be a great help.

Global warming is a catastrophe in the making. If you read sarcasm into the statement, that is fine; if you take it seriously, that is OK, too. The treatment of global warming here is not intended to prove or discredit the phenomenon. My purpose is to cast the bright light of practical environmentalism on this really, really big environmental issue and use the pillars to examine possible responses to the perceived crisis.

Remember that the pillars are useful when considering actions, policies, and programs. The pillars will not tell you if global warming is real or not, but they will help us examine our response and judge if that is worthwhile. There are many potential responses that have been proposed to prevent the coming catastrophe. They generally revolve around how we use and transform fossil energy—coal, gas, and oil, mainly. Energy transformation is the key because the main greenhouse gas, CO_2, is generated as a significant by-product of transforming fuel into electricity, heat, or power. We all are partially responsible to the degree that we depend upon fossil fuels to operate our electric devices, heat and air condition our buildings, and transport us from one place to another.

Let us take a brief and basic look at how carbon makes it way from being bound within fossil fuel to being liberated into the atmosphere. Carbon (C) in the fuel combines with oxygen (O_2) in the air to form carbon dioxide (CO_2).

$$C + O_2 = CO_2$$

This simple chemical reaction occurs every time you start a gas-powered engine, every moment that a coal-burning power plant is on line, and every time your natural gas furnace or water heater kicks on at home. It also occurs with every breath we take, or that any animal takes for that matter. Humans, and mammals in general, consume oxygen and expel CO_2. Luckily, plants do the exact opposite and consume CO_2 and produce oxygen. CO_2 is an intimate inextricable part of the natural life cycle.

Many of the proposed "solutions" to the global warming problem have very big price tags. One of the reasons for this is because they treat CO_2 as a pollutant instead of a by-product. A pollutant is something that can be caught, trapped, and contained. It is generally a small fraction of what is emitted by a process and not an integral part of the process itself. Pollutants are undesirable irritants that we would generally prefer to eliminate. Zero pollution is good by definition. CO_2, however, does not fit this definition, even considering that more and more people are labeling it a pollutant. Reducing CO_2 to zero is not only foolishly impractical but a very, very bad idea.

Let us look at a few of the most popular proposals and see what we get when we run them through the pillars.

One popular and pricey idea is that of carbon sequestration. This concept is basically to catch and store the CO_2 you get from burning fossil fuels before it ever reaches the atmosphere. This idea is usually directed toward coal-burning electric power plants.

Early designs envision some kind of chemical or physical process that is added on to the back end of a power plant to remove the CO_2 from the smokestack. Once you have caught the CO_2, you send it through a pipeline and then inject it deep underground, hopefully never to be seen again. Coal-fired power plants are attractive to sequestration proponents because coal produces more greenhouse gases than other fossil fuels and because they represent a large and concentrated source of CO_2 emissions.

Carbon sequestration is expensive for two main reasons. One is that the CO_2 removal equipment itself is big simply due to the size of the power plants themselves. This is to be expected as most process equipment in a coal-fired power plant is big and expensive. The other reason, however, is a bit more subtle. Some carbon sequestration designs, and the one we will focus on here, require a lot of energy to operate, up to 33% of the energy that is produced by the power plant. There is also a significant energy input needed to transport the CO_2 to the disposal site. This implies that existing coal-fired power plants will be significantly derated and brand-new power plants will be needed to make up for the shortfall in capacity.

The other issue, of course, is this: just where do you "dispose" of the CO_2? Carbon sequestration assumes that it is possible and safe to inject CO_2 gas deep underground in a pressurized state. This is already being done in oil fields as a way to improve the recovery of the oil itself, but there is a good bit of debate around this issue as a pure disposal method. People worry about the potential of the CO_2 leaking back up to the surface and creating a hazardous local environment.

Carbon sequestration is a brand-new technology and really deserves far more description than I have given here, but I hope at least the basics are clear. Let us proceed with the pillars and see if a more detailed analysis is warranted (Table 12.1).

Environmental degradation gets a zero in my analysis. It may seem strange that it does not receive a positive score when the whole point of carbon sequestration is to remove the "pollutant" CO_2 from the atmosphere. There are two main environmental issues that arise, however, that cause the zero result. First, carbon sequestration removes CO_2 from the atmosphere but then deposits it in the ground. It is a classic pollution transplant problem. Second, and possibly more important, is that the additional power requirements for the carbon sequestration process also generate additional traditional air pollutants. The type and quantity of pollutants generated would depend upon what types of new power plants are built to provide the additional power needed to run the carbon sequestration processes. If all this new power

TABLE 12.1
Sequester Carbon Dioxide

Pillar	Score
Environmental degradation	0
Resource conservation	−2
Economic progress	−2
Personal benefit	0
Total	−4
Indication	Do not bury the carbon

were renewably based, such as solar, wind, geothermal, and so forth, then perhaps this additional pollution load would be insignificant. This is a very unlikely scenario because the amount of additional power required could be very, very large. As mentioned previously in Chapter 6, half of the electricity generated in the United States comes from coal-burning power plants. Equipping all these plants with carbon sequestration technology would be a massive undertaking and could require many new power plants to be built. At this time it is difficult to see how this could be done without relying on traditional fossil fuel power plants. So the perceived environmental benefit of reducing CO_2 emissions is tempered by environmental issues with CO_2 separation and storage. We are left with an environmental draw in this case, and the environmental degradation pillar reflects this with the zero score.

Resource conservation gets a score of −2. This is due to all the additional power required for the carbon sequestration process. Lots of additional power consumes lots of resources. Even if renewable resources are used, they would be dedicated to the carbon sequestration operations instead of being dispatched to meet the needs of our homes, offices, and factories. If we accept the statistics that coal-fired power plants supply half of all US electricity needs, and that carbon sequestration equipment will require one-third of the power produced by these plants, then carbon sequestration will require an increase in power generation of approximately 16% across the United States ($0.5 \times 0.333 = 0.167$). This is a very large increase in energy production, basically requiring a new power plant to be built for every six existing power plants. The best-case scenario to meet these power needs would be through a combination of renewable energy and natural-gas-fired generation. Even in this case the demand for fossil fuels would increase and the promise of renewable energy would be diverted toward environmental cleanup instead of fossil fuel reduction. This results in the strongly negative score for the resource conservation pillar.

Economic progress also gets a −2 score. As envisioned here, carbon sequestration is a huge economic burden and imposes large inefficiencies on the electric utility system. Electricity prices could see very large increases without any significant economic benefit to our society. Your electric bill at home would go up, perhaps by 15%–20%. So would the electric bills of our schools, offices, and factories. Certainly some engineering and construction companies would be very busy but at a large cost to the rest of our economy in the form of strong inflationary pressure on many of the goods and services we consume.

Personal benefit gets a zero, and I think this is very generous. The biggest individual detriment will be through higher utility costs. Electricity would certainly be more expensive, but natural gas prices could rise as well if much of the new power required for carbon sequestration is generated from natural-gas-burning power plants. It is conceivable that it could cost a lot more to heat and power our homes. This negative impact may be balanced by a perceived benefit of avoiding the global warming catastrophe. This, of course, assumes that an individual believes in the global warming predictions and values actions that reduce the likelihood or magnitude of the predicted changes. I personally would score this pillar as a −1 or even −2, but that depends on your personal values and priorities. I scored it a zero here assuming the perspective of someone who is very concerned about the potential negative consequences of global warming. This balances the negative impacts of increased costs in general and results in a zero score for the pillar.

In total, carbon sequestration gets a score of −4. This is a "Heck No!" result as far as the pillars are concerned. This shows that the perceived environmental benefits are far outweighed by practical environmental, resource, economic, and personal concerns. This result should be a "wake-up call" to all those who are actively planning to pursue carbon sequestration as a global warming mitigation strategy. Carbon sequestration systems that come with high parasitic energy requirements simply do not make sense from the practical environmentalism perspective. This pillar analysis tells us that we should look for an improved carbon sequestration process or look for entirely different means to address CO_2 concerns.

One such entirely different approach to solving the global warming "problem" would be to institute renewable energy portfolio standards. These standards would be a government mandate, either at the federal or state level, that a certain percentage of our total electricity be generated by renewable energy technologies. These standards would likely be directed toward the electric utility companies, and they would be required to demonstrate that part of their power production assets, in essence their energy supply portfolio, comes from windmills, solar cells, hydroelectric dams, biomass power plants, and so forth. In fact, renewable energy portfolio standards already exist in many jurisdictions.

Renewable energy standards force utility companies to reduce the amount of fossil fuel they consume by making them invest in and employ renewable energy technologies. Reducing fossil fuel consumption reduces greenhouse gases. Higher renewable energy standards result in greater reductions in greenhouse gases emitted. Let us see what the pillars say about this approach (Table 12.2).

Environmental degradation gets a score of +2. Encouraging renewable energy generation is certainly expected to yield environmental benefits via less air pollution. Global warming proponents would also value the reductions in CO_2 from fossil fuel combustion. Renewable energy portfolio standards accelerate the application of new energy technologies with fewer emissions and lesser environmental impact that replaces old technology with greater emissions and greater environmental impact. While even renewable technologies can have some negative environmental impacts, they are typically minor, and this does not preclude renewable energy portfolio standards from receiving a +2 score in my view.

TABLE 12.2
Renewable Energy Portfolio Standards

Pillar	Score
Environmental degradation	+2
Resource conservation	+2
Economic progress	−1
Personal benefit	+1
Total	+4
Indication	Renewable is where it is at!

Resource conservation gets a +2 score. This is a direct result of the targeted reduction in fossil fuel consumption. Renewable energy portfolio standards, set at a significant level, could have a marked effect on how much coal, gas, and oil is consumed by power generation facilities.

Economic progress receives a −1 score in this analysis, basically because renewable energy portfolio standards would force utility companies to choose higher-cost technologies to produce power. This is obviously economically inefficient, at least in the short term, and therefore merits the negative result for this pillar. The standards do permit some choice, however, in just how utilities would comply with the standards, and so, it is reasonable to assume that utilities could optimize the process and choose lower-cost options of all the available renewable energy technologies. This ability to find a minimum cost approach to meet the standards mitigates the negative impact to this pillar and keeps it at a −1 instead of a −2.

Personal benefit gets a +1 score. Traditional air pollutants as well as greenhouse gases would be reduced. Our power generation and distribution system would likely become more diverse and less vulnerable to commodity price swings of fossil fuels. A number of incentives could be offered by utility companies to individual customers to install renewable energy technologies at homes and businesses that could lead to reduced energy bills. A number of new industries and businesses could be created to support the widespread installation of a broad range of renewable technologies. However, there would likely be at least a modest increase in utility bills for many consumers. Forcing a nonoptimum economic solution is expected to result in higher prices. Considering all these likely impacts, it seems that renewable energy portfolio standards would result in a net benefit to individuals. There are a few potential drawbacks, and these keep us from giving this pillar a +2 score. If you are very sensitive to utility costs, you might disagree with my weighting of factors and prefer a score of zero.

The total score for all the pillars results in +4, a very positive result. Again, if your personal benefit score was more negative than mine, you might come up with a +3 score, still a positive result. Even if you think global warming is overhyped nonsense, it still may make sense to support renewable energy portfolio standards. These standards could reduce fossil fuel consumption, which tends to extend reserves and puts downward pressure on prices. The standards could also reduce emissions of traditional pollutants from fossil-fuel-fired power plants. Even discounting the contribution of the +2 environmental degradation score, the pillars still show a positive total result.

What a large contrast compared to the carbon sequestration approach—+3/+4 versus −4! Even though both approaches have a similar desired outcome of CO_2 reduction, the pillars show vastly different scores depending upon the path selected. A number of cliché sayings leap to mind such as, "the devil is in the details," or "the end does not justify the means." These vastly different scores again showcase the value of the pillars in helping us focus on practical environmental action that offers widespread benefit to individuals and society.

Let us look at one more popular approach to solving the global warming crisis. Many have proposed that we should institute a carbon tax to reduce the amount of carbon emitted to the atmosphere. This tax would be applied directly to fuel

TABLE 12.3
Carbon Tax

Pillar	Score
Environmental degradation	+1
Resource conservation	+2
Economic progress	−2
Personal benefit	0
Total	+1
Indication	Proceed with caution

purchases, be it gasoline for automobiles, coal consumed by power plants, or natural gas flowing to industry or heating our homes. The goal of taxing carbon is to make it more expensive to burn fossil fuels, thereby encouraging increased use of renewable energy sources and/or energy conservation in general. Fuels with higher carbon content would be taxed more heavily than fuels with less carbon. All this would hopefully lead to less CO_2 emitted to the atmosphere and less chance for the global warming catastrophe to occur. Let us run carbon taxation through the pillars and see what we get (Table 12.3).

Environmental degradation gets a score of +1 for many of the same reasons that renewable energy portfolio standards received a score of +2 for this pillar. The primary difference warranting a slightly lower score is that carbon taxation does not necessarily force the widespread adoption of renewable energy technologies. Carbon taxation encourages renewable energy, but only indirectly by changing the purchase price of fuels. As before, encouraging renewable energy generation is expected to yield environmental benefits via reduction in pollution from burning fossil fuels.

Resource conservation gets a +2 score for a couple of reasons. First, we would expect to see greater use of renewable energy as fossil fuels become more expensive. Second, we would expect to see improvements in efficiency in the use of fossil fuels. When fuel is very expensive, we tend to optimize and minimize our needs to get the most out of the fuel itself. If we think about our own personal transportation needs, a carbon tax on gasoline might prompt us to drive a more fuel-efficient car, or buy an electric car that we could charge with solar panels, or even ride the bus from point A to point B. Carbon taxation gets at both the supply side and the demand side of fossil fuel use and deserves the highest score for this pillar.

Economic progress receives a −2 score in this analysis, basically because carbon taxation does not guarantee widespread adoption of renewable energy technology, and it imposes additional costs on virtually every aspect of our economy. Carbon taxation arbitrarily changes the economic balance among energy technologies and imposes economic penalties on some industries that are less able to adapt to renewable technologies. It extracts its penalty up front and diminishes our ability to optimize the balance between renewable and fossil fuel technologies. These economic inefficiencies seem greater to me than for the renewable energy portfolio standards and thus lead to the more negative score for this pillar.

Personal benefit gets a zero score. As with renewable energy portfolio standards, traditional air pollutants and greenhouse gases would be reduced, and our power generation and distribution system would likely become more diverse. I would expect, however, that energy cost increases could be more significant for many consumers under a carbon tax plan. I think this is particularly true for consumers who do not have the economic power to invest in new technologies. They are the ones who could not afford a new fuel-efficient car and who are forced to drive an old gas guzzler if they drive at all. They are the ones who could not buy a new furnace for their homes and would be more vulnerable to increasing utility costs. Basically, a carbon tax could become a very regressive tax and be burdensome to those at the lower end of our economic spectrum. Considering all these likely impacts, it seems that carbon taxation would be a wash at best, and therefore gets a zero score. Again, if you are very sensitive to utility costs, you might disagree with my weighting of factors and prefer a score of −1.

The total score for all the pillars results in +1, a barely positive result. Again, if your personal benefit score was more negative than mine, you might come up with a zero score, and an ambiguous result. This zero or near-zero result should be of concern both to individuals and policy makers. The pillars indicate that carbon taxation is probably not a good idea, especially considering that other global warming mitigation measures score much higher.

We have briefly discussed three possible approaches to reducing CO_2 and heading off the global warming problem. Recognize, however, that there are many other potential solutions to this issue, some of which could be significantly better and of greater value than the three described here. I chose carbon sequestration, renewable energy portfolio standards, and carbon taxation simply because they are often discussed in the current debate about how we should respond to the global warming issue. Hopefully, their inclusion here illustrates that the pillars can be applied successfully to the biggest environmental issue of our day, and that the pillars can easily differentiate between courses of action.

The pillars have shown us that these three approaches have significantly different results as viewed by a practical environmentalist. So, regardless of your acceptance or contempt of the global warming theory, you can use the pillars to analyze some proposed "solution" and determine if it is generally beneficial or detrimental. I suggest that this be done whether you are a true believer or a skeptic. Choosing wise courses of action is critically important when we are dealing with such a huge, expensive, all-encompassing issue. If you are a skeptic, there is great value in not squandering our resources on ill-fated solutions. If you are a believer, there is great value in investing our resources to achieve the greatest impact. Either way, the pillars can help us see our way to a better future.

REFERENCES

Le Treut, H., Somerville, R., Cubasch, U., Ding, Y., Mauritzen, C., Mokssit, A., Peterson, T., and Prather, M. 2007. Historical overview of climate change. In: *Climate Change 2007: The Physical Science Basis. Contribution of Working Group I to the Fourth Assessment Report of the Intergovernmental Panel on Climate Change.* Solomon, S., Qin, D., Manning, M., Chen, Z., Marquis, M., Averyt, K.B., Tignor, M., and Miller, H.L. (eds.). Cambridge University Press, Cambridge, U.K.

Broecker, W.S. 1975. Climate change: Are we on the brink of a pronounced global warming? *Science* 189:460–463.

IPCC—Intergovernmental Panel on Climate Change. 2010. http://www.ipcc.ch/ (accessed September 10, 2010)

Tans, P. 2010. Mauna Loa Monthly CO_2 Concentration Dataset. National Oceanic and Atmospheric Administration Earth System Research Laboratory. http://www.esrl.noaa.gov/gmd/ccgg/trends/ (accessed September 26, 2010).

Hansen, J.E., Ruedy, R., Sato, M., and Lo, K. 2010. NASA GISS Surface Temperature (GISTEMP) analysis. In *Trends: A Compendium of Data on Global Change.* Carbon Dioxide Information Analysis Center, Oak Ridge National Laboratory, U.S. Department of Energy, Oak Ridge, TN. doi: 10.3334/CDIAC/cli.001.

IPCC—Intergovernmental Panel on Climate Change. 2001. Graph of Twenty Glaciers in Retreat Worldwide. http://www.grida.no/publications/other/ipcc_tar/?src=/climate/ipcc_tar/wg1/fig2-18.htm (accessed September 10, 2010)

Davis, C.C. 2010. Does Climate Change Promote Invasive Species? http://harvardmagazine.com/harvard-in-the-news/climate-change-benefits-invasive-species (accessed September 13, 2010).

NOAA—National Oceanic and Atmospheric Administration. 2010. December Global Ocean Temperature Second Warmest on Record. http://www.noaanews.noaa.gov/stories2010/20100121_globalstats.html (accessed September 10, 2010).

13 Other Measures of Environmental Performance

As environmentalism has become more and more important over the past few decades, a number of methods have arisen to chart our progress toward a healthier planet and a better future. These methods have generally been intended to be a measuring stick that informs us of our impact upon the earth. This chapter is devoted to a discussion of some of these methods and a comparative analysis with the Practical Environmentalism approach to bettering our world. The number and variety of these measures are a testament to our societies' burgeoning acceptance of environmentalism as a basic component of our culture. The following list of environmental quality metrics is certainly not complete but hopefully long enough to grant us a decent picture of the various attempts we make to quantify and qualify our impact upon the natural world.

- Carbon Footprint
- Carrying Capacity
- Life-Cycle costing
- Government & Scientific Reports
- Green Accounting (ISEW)
- Simple Monetary Economics

CARBON FOOTPRINT

In the past decade or so, the concept of a carbon footprint has become a well-publicized measure of environmental impact. We touched upon it in the last chapter regarding global warming and global climate change. Again, our carbon footprint is a measure of how much our lifestyles rely on the energy contained within fossil fuels. As our own footprints show evidence of where we have physically been, our carbon footprint shows evidence of how much fossil fuel we have directly or indirectly consumed.

Our carbon footprint tries to measure the carbon dioxide (CO_2) produced by our existence. For example, it seeks to link us to the CO_2 from growing the food that we purchase in the grocery store. Part of the carbon present in the fuel for the farmer's tractor or combine is attached to us. Burning this fuel powered the equipment that the farmer used to grow the crops that we eventually consumed. The carbon present in the fuel that the farmer used was converted into CO_2 and emitted into the air, adding to the stock of atmospheric CO_2 being blamed for changing our climate.

This, of course, is only the beginning of this accounting exercise. The trucks that transport the farmer's crops to storage silos, warehouses, and distribution centers use fuel as well and liberate CO_2 into the atmosphere. Food manufacturers use energy and generate CO_2 to process the crops they purchase from the farmer and turn them into the prepackaged prepared foods that we store in our pantry. Even the packaging materials used add to our carbon footprint. The tin can that holds our soup was manufactured and transported to the soup factory at a cost of CO_2 liberated into the atmosphere. The paperboard box that holds our breakfast cereal adds carbon, too. Every time we pop a prepared meal into our microwave oven, we are benefiting from agricultural, manufacturing, and transportation processes that have a cost of carbon associated with them and then linked to us.

As you can see, even the simple task of popping some popcorn to enjoy while watching our favorite TV show has a complicated and detailed set of calculations to account for the carbon released into the environment. But do not forget the carbon associated with your favorite TV show either. It takes electricity to power your TV, and that electricity could easily have been generated by combusting fossil fuels. Also remember that producing your favorite TV show requires energy, too. Energy is needed to construct the sets at the studio, run the cameras, power the lights, and even to make the cast and crew members' lunch.

Of course, we have only begun to account for the carbon associated with our lives. To accurately measure our carbon footprint, we have to account for the carbon emissions associated with all the items we buy and all the things we do. This headache-generating exercise is rather impractical and something I would rather avoid. It appears to be a never-ending backward examination tracing every item that touches our lives back to its beginnings in the earth. Yikes!

This tedious calculation is rarely employed. As you can see from the preceding few paragraphs, an individual's carbon footprint is essentially incalculable from a bottom-up approach. Tracking our carbon backward is simply unworkable. Now, I imagine some people have tried. If I was teaching an environmental studies course, I just might require this of my students as a class project. We would soon learn that we have to put some limits on the exercise or be willing to make some gross assumptions to complete the calculations. We could reasonably account for some of the bigger contributions to our carbon footprint by checking our utility bills or estimating how many miles we drive in a year and calculate the amount of gasoline our car requires to travel this distance. I imagine this would be a fine educational endeavor and perhaps open our minds to a new appreciation of the energy and carbon cost of things we often take for granted. However, students have little choice but to comply with their professors' expectations if they want to earn a decent grade. I wonder how many would undertake this exercise of their own free will. How many of us postgraduates would undertake such an exercise?

Still you can do a little research and learn what your carbon footprint is. There are scientific studies, government reports, and computer programs that will provide this information. They are not based on your individual situation, however. They are not the detailed bottom-up approach we discussed earlier but rather represent a top-down approach. Our individual carbon footprint is typically inferred from the carbon footprint of our nation. It is at the national level where we can reasonably gather the energy usage data we need. As a nation we can keep track of the amount of coal we mine and the number of gallons of oil we draw from the ground. We can also calculate the amount of CO_2 generated from each ton of coal or barrel of oil that we consume. Thus we know our national carbon footprint in general terms. We can divide this number by our population and arrive at a "calculated" individual carbon footprint. It has only the most tenuous relation to our individual circumstance, but in aggregate it is reasonably accurate for all of us. It tells us how big an American's energy appetite is compared to that of someone from France, or Kenya or New Zealand. In this case, though, an overly zealous appetite bypasses our big bellies and goes straight to our feet. We do not need to loosen the belt buckle as much as we need to get a bigger pair of shoes.

This top-down calculation can be modified somewhat to generate factors that can be used at the individual level. We could disaggregate our energy use into industry segments, for example. We could estimate how much energy the aluminum industry uses, or the paper industry or agriculture in general. We could assign energy use by industry to the products they make and then estimate the carbon associated with their products. Then, if we knew how many cans of soda we consumed in a year's time, we could reasonably estimate the part of our carbon footprint associated with feeding our soda pop habit.

We could also undertake regional comparisons and calculate how much energy is used for air-conditioning homes in Florida versus homes in Minnesota. We could use this information to estimate an individual's carbon footprint for cooling his or her home depending upon where he or she lived. Thus we could begin to tailor our national statistics to produce local factors that we could use to better estimate our individual carbon footprint.

So, how do we employ this information? Now that we have an estimate for our carbon footprint, and assuming we are willing to accept the inaccuracy present in it, just what do we do with it? Is it meant to motivate us big-footed Americans? Should we strive to reduce our carbon footprint to be on par with our European friends? Should we reduce it even further than that? The environmentalist would answer these questions with a resounding YES, of course, but it is an overly simple question and an overly simple answer. The carbon footprint method is a single-issue metric. All it cares for is CO_2 liberation, primarily through fossil fuel energy use, and all it can tell us is that we should undertake action to make our carbon footprint smaller. It advocates for resource conservation but ignores other types of environmental impacts as well as economic and social concerns. While the carbon footprint metric is an interesting addition to the global climate change issue, it is rather limiting in its usefulness as a practical tool we can use to foster sustainable environmental progress from an individual's perspective.

CARRYING CAPACITY

This concept represents a theoretical linkage between population and environmental degradation. The basic idea is that our environment, essentially the earth itself, can only support so many people. The carrying capacity is the number of people that the earth can support in a sustainable manner.

Here again is the concept of sustainability. We just cannot seem to get too far away from it when discussing environmental matters. Carrying capacity implicitly includes the idea of sustainability. Our earth can only provide resources year after year for a certain number of people. The earth has a limited capacity for growing food, for regenerating clean water, for producing renewable energy, and generally for making resources available to humanity that we desire. The carrying capacity concept is based upon these natural limits as well as how quickly humanity consumes resources. If each of us lives very frugally, the earth can support many. But if each of us desires to live lavishly, the earth can only support a relative few.

The carrying capacity concept appeared in the early 1900s in the fields of livestock and wildlife management (Sampson 1923). It was initially applied to range management studies where it was desired to quantify how many animals could graze on a certain range without degrading the range itself and diminishing the productivity of the natural plants and grasses that grew there. It explicitly made the connection between the productivity of plant species and the demands put upon them by grazing animals. This application of carrying capacity also finds a place within Garrett Hardin's essay "Tragedy of the Commons," which used a publicly held and shared pasture as a metaphorical example to raise concern about population growth and natural resource consumption outpacing nature's capability to provide for all of us (Hardin and Baden 1977).

Hardin's concerns are the basis for applying carrying capacity to environmentalism. Carrying capacity is also easily included within the anticonsumption ethic discussed in Chapter 3. Recall the I = PAT equation that related environmental impact to population, affluence, and technology. Carrying capacity asserts that I, environmental impact, reaches a maximum tolerable limit at some point. This limiting value is the earth's carrying capacity. At this point nature is barely able to keep up with the demands we place on her. If we increase our impact, through population growth or increased consumption (affluence), we will ultimately diminish nature's capacity to provide resources to us.

Carrying capacity is a very interesting theoretical concept. It finds a comfortable home within academia and is applied in many fields. It is often included within academic and scientific literature. It is difficult, however, to apply the concept to environmentalism on a practical level. One of its problems revolves around the issue of time. Carrying capacity depends upon the state and productivity of nature. If our environment is healthy and unstressed, it can produce more resources than if it is degraded. A mature mixed-species forest provides habitat for wildlife, wood fuel for some level of renewable use, watershed protection, soil erosion protection, and even temperature moderation. If we clear-cut this forest, many of these provisions vanish or are at least significantly reduced. Thus, the carrying capacity can change from one year to the next as human actions alter the environment. So, how would we go about calculating the earth's carrying capacity? How do we really know how many people the earth can support?

These are both tough questions. If you search through the scientific literature on this topic, you will likely find a wealth of opinions but no consensus as to the number itself. This makes it difficult to use carrying capacity as a metric to help us decide what action we should take to benefit the environment. Even if we ignore the ability for carrying capacity to change, the real measure of interest for environmental decision making is how our current impact (PAT) compares to the earth's carrying capacity. Some would propose that we already exceed the earth's carrying capacity, that our current population, affluence, and technology have already degraded our environment such that it is less productive. They would say that we need to reduce population or affluence, or improve technology. Others might scoff at this notion and assert that the earth continues to provide all the resources that we require. Carrying capacity tends to divide the Anticonsumption ethic from the Cornucopian ethic. Proponents of both could give evidence to support their views. In this respect, carrying capacity becomes more a point of debate than a useful tool to encourage collaboration and inspire environmental progress.

Carrying capacity is not very helpful on an individual level either. Much like the carbon footprint metric, carrying capacity essentially states that reducing our impact upon the world is desirable. We already knew that though, right? In its favor, however, it is a more complete metric in that it easily considers any issue that may impact the environment. Where our carbon footprint is basically a single-issue metric focused on CO_2 and global climate change, carrying capacity is relevant to any issue impacting the earth's capability to support humanity. Water supply, hazardous waste, and endangered species issues do not easily fall within the carbon footprint approach but can all be analyzed under the carrying capacity metric.

Drawing comparisons between our individual actions and the carrying capacity of the earth would be a tenuous and tedious exercise at best. The metric is so broad that it is difficult to see if what we do individually has any impact on it whatsoever. In terms of environmentalism, its best use probably remains in academic discourse and as a general motivating factor to remind us that our existence does impact our environment and there just might be some limits to the environment's ability to stand our existence.

LIFE-CYCLE COSTING

This approach attempts to include all the costs incurred by the purchase, use, and disposal of a product or service over its lifetime into the purchase price of the product. From an environmental perspective, this represents a way to include the costs of negative externalities into the product or service at the time the purchasing decision is made. As mentioned briefly in Chapter 8, negative externalities is a term used in economics to describe the unplanned or unaccounted-for harmful side effects from an activity. Externalities can be beneficial as well; we would label them as positive externalities, but for our purposes they are typically negative and represent unwanted environmental degradation or natural resource depletion. These externalities may be actual monetary costs that are borne in some fashion other than the transaction price of the good, or they may be social or environmental costs that are not reflected in monetary terms.

For example, many of the products we purchase end up in our trash cans to be hauled to our local landfill. Most of us pay for the privilege of having someone take our garbage away from our homes. So, each item that we purchase and then throw away has an additional cost for disposal that is not included in the purchase price. In my hometown we pay for garbage collection in two ways. There is a monthly fee that is attached to our water bill, and we also purchase garbage stickers for a couple of dollars each. Each standard garbage can that we place on the curb requires a sticker in order for the garbage collector to dump it in the truck and haul it away. I can fill up the garbage can for $2. Everything that I put in the can then has an additional cost proportional to the volume it occupies in the can. If a single item takes up half the can, it costs me a dollar to get rid of it. Every pizza box, candy wrapper, old pair of shoes, or watermelon husk that I toss in my trash can has some fraction of the $2 disposal cost that is unaccounted for in its purchase price.

Say, for example, that I purchased an inflatable pool for my children to use during the hot Iowa summer. It is a pretty big one because I have four kids and they all like to cool off in the water. Let us also say that, after a season's use, one of my son's friends accidentally punctured the wall of the pool, and during the excitement of all the kids bailing out of a collapsing pool, this puncture turned into a major gash in the side of the pool. This gash is long and jagged and well beyond my capability to repair it. My pool has suddenly become garbage, and will cost me an extra $2 assuming that I can shove it into a single trash can. Life-cycle costing would maintain that this $2 disposal cost should be included in the initial purchase price. I should be aware of how much it will cost to dispose of the pool when I am done with it. If total costs were generally included in purchase prices, maybe I would make different purchasing decisions, perhaps better purchasing decisions.

Think what might happen if landfills and garbage collection services did not exist and we had to figure out how to dispose of our refuse on our own. Maybe we would dig a pit and bury it in our backyard. Perhaps we would throw it in a heap and light it on fire. Or, we could toss it in the local river and let it become someone else's problem downstream. Maybe, though, we would change our behavior such that we only bought things that were 100% recyclable or compostable, including the packaging. Is this so unrealistic?

While disposal costs are a good example of life-cycle costing, including environmental impacts that are sometimes forgotten when we stand in the checkout line at the store, the life-cycle costing method can include much more than that. Life-cycle costing is interested in the total costs of producing the product that I eventually purchase. We know many of these costs already. Manufacturers know the costs of their raw materials, the cost of the labor to produce the product, the cost of the energy required to manufacture the product, and the cost to transport the product from the factory to the store where I can go to purchase it. These costs are easy to calculate in monetary terms and are already part of the purchase price of the product. Cost accountants can very easily figure these things out to the nearest penny. What is missing though are the hidden costs. Costs of pollution or resource depletion or even government subsidy are not easily calculated, or are deliberately kept separate. These costs are associated with the product I buy, but I do not pay them directly. The costs of the air pollution emitted from the factory's smokestacks are borne by the people living downwind, perhaps as increased healthcare costs associated with air that is less clean. The costs of depleting a groundwater aquifer to supply the factory are borne by the nearby residents who also depend upon the aquifer for their water, perhaps in the cost of drilling deeper wells or connecting to a municipal water supply. The costs of the local government tax breaks to entice the factory to remain where it is instead of relocating elsewhere are borne by local taxpayers, perhaps in higher property taxes. None of these costs, and quite possibly many others, are paid by me at the cash register where I buy my pool.

Faithfully employing life-cycle costing suggests that environmental degradation would tend to decrease. This would happen out of pure self-interest on the part of the consumer. Products that came to us with high costs to the environment or society would see price increases and could be disadvantaged in the marketplace. No more could environmentally destructive products hide behind a low-priced façade. Not only would they simply be more expensive but we might also realize the toll they exact in environmental damage and prefer to find an alternative.

Much like carbon footprints and carrying capacity, life-cycle costing suffers a great need for data and lengthy detailed calculations to be an accurate environmental measure. It is very difficult to reach back too far in the supply chain to find all the hidden costs that are associated with each product. Manufacturers do not have much incentive to provide this information. It essentially makes any product look worse. Even environmentally friendly products can have hidden costs. By definition these hidden costs are less than those of more toxic products, but they still can exist. Government regulations would probably be required to enact life-cycle costing at the consumer products level, and much oversight and administrative bureaucracy would probably be required to ensure reasonable accuracy and fairness. I am not personally convinced that this would result in a net environmental benefit.

Alternatively, if we wished, we could individually incorporate some of the components of life-cycle costing into our purchasing decisions. With relative ease we could estimate the disposal costs of things we buy. We could consider the ability to recycle or compost packaging material as we consider the product on the shelf, assuming that recycling and composting are essentially free. We could investigate what materials go into the products we buy and determine whether any of these materials are associated with toxic compounds. This could be rather complex and would require some time to learn about the toxicity of common materials used in manufacturing processes. There are lots of manufacturing processes after all. Finding a good book on the subject or subscribing to a couple of relevant magazines would go a long way in enlightening ourselves as to which materials should trigger our concern. Or we could search out which manufacturers are the greatest polluters and avoid the products they make. This might be somewhat simpler than becoming educated on material toxicity. Manufacturers are required in many cases to report on the amount of pollution they emit in the form of air and water emissions and hazardous waste. Some advocacy organizations keep close tabs on this information and summarize it for publication. A quick Internet search would reveal a number of "dirty dozen" lists of corporate polluters, as well as lists of elected officials who are perceived as supporting the status quo of rules and regulations that allow this pollution to continue. The phrase "dirty dozen" is very popular in this context and has become ingrained in the American vocabulary, probably in homage to the 1967 film about World War II and its cast of 12 unscrupulous characters.

Much like carrying capacity, life-cycle costing is a multi-issue metric and can include many aspects of our modern environmentalism. Determining appropriate monetary costs for hidden social and environmental factors may prove difficult, but the life-cycle costing method itself is robust enough to include these considerations. Theoretically at least, it is broad enough and flexible enough to serve as a yardstick for monitoring environmental performance. As mentioned in the previous paragraph, while we cannot employ life-cycle costing with complete accuracy, we can implement its logic and approach for our individual decision making and at least give some partial measure of due consideration to environmental factors.

GOVERNMENT AND SCIENTIFIC REPORTS

There are a vast number of environmental reports published by academic, advocacy, government, and scientific sources. These are the "experts" that we often entrust to explain environmental issues to us and to provide solutions to environmental problems. Some reports are very specific and concentrate on a particular environmental issue in a particular place. Others take a very broad view and attempt to qualify the health of our global environment in general. Reports published by advocacy groups and government bodies are probably the most common that we will encounter. This information is often readily available from Web sites, and many groups have a lengthy list of publications. Academic and scientific research is much less widely distributed and tends to remain within scholarly publications that most of us have little reason to peruse. This research, however, is often the foundation for advocacy and government reporting and, as such, is critically important in creating knowledge and understanding. Here are just a few examples of the kinds of reports that are available regarding the health of the environment.

The Worldwatch Institute is a research-based advocacy group that has published an annual report on environmental quality for many years (Worldwatch Institute 2010). If you visit its Web site, you will see that the institute has been around since 1974. It played a part in the emergence of environmental awareness that occurred in the late 1960s and early 1970s. Worldwatch's focus is global, much as its name implies, and revolves around three basic areas. These areas are climate and energy, food and agriculture, and the green economy.

I chose to include Worldwatch here specifically because of its more than 25-year history in reporting on environmental quality on a global basis. Its reports read like a book with referenced chapters, discussing particular environmental issues or threats that are deemed central to the health of the environment at the time. For example, the report for 2010, "State of the World 2010: Transforming Cultures: From Consumerism to Sustainability," obviously has a central theme related to transitioning our society toward sustainability. The 2009 report, "State of the World 2009: Into a Warming World," placed special emphasis on global climate change. These reports, and the Worldwatch Institute itself, are a good example of an advocacy organization taking on the challenge of reporting on the health of the global environment.

Another organization willing to tackle reporting on the health of the global environment is the United Nations, specifically the United Nations Environment Programme (UNEP). This quasi-governmental organization, sponsored by national governments, is naturally suited and positioned to take on global environmental issues. The UNEP has published a series of yearbooks, much like the Worldwatch Institute, that try to raise awareness of the current state of the environment on an annual basis (UNEP 2010).

For example, UNEP's 2010 yearbook is composed of six chapters that mirror UNEP's six main focus areas. They are

1. Environmental Governance
2. Ecosystem Management
3. Harmful Substance and Hazardous Waste
4. Climate Change
5. Disasters and Conflicts
6. Resource Efficiency

The 2010 Yearbook also has a central theme regarding water that is interwoven through each chapter. Naming this report a "yearbook" is very appropriate as it focuses at times on notable events of the past year within each chapter. It is an interesting combination of journalism and environmental science. As with the Worldwatch Institute's annual reports, the UNEP yearbooks are very informative in a very general and very readable form.

Some national governments also accept the challenge of reporting on the health of the environment within their borders. In the United States, the Council on Environmental Quality (CEQ) has historically attempted to report on the health of the American environment. The CEQ was formed by the National Environmental Policy Act of 1969 and produced a series of environmental quality reports, often called CEQ annual reports, from 1970 through 1997. Unfortunately for those of us who appreciate environmental data and review, 1997 was the last year of the CEQ annual report, it having fallen victim to a congressional paperwork reduction act that eliminated many US government reports from publication.

The CEQ annual reports were multichapter books as well. They were focused on the United States, of course, but did have some references to the global environment, including a full chapter devoted to international cooperation to address global environmental issues in the 1997 annual report. The CEQ annual reports showed selected trends in some of the issues they discussed, enabling the reader to easily get a sense if a particular environmental indicator was improving or worsening. The last annual report in 1997 was composed of the following chapters:

1. Population
2. Economy and Environment
3. Public Lands and Recreation
4. Ecosystems and Biodiversity
5. Air Quality
6. Aquatic Resources
7. Terrestrial Resources
8. Pollution Prevention, Recycling, Toxics, and Waste
9. Energy

10. Transportation
11. Global Environment

The later editions of the CEQ annual report also contained information about NEPA itself and selected NEPA legal cases. Most editions of the report were well supported with data tables and information.

The Council on Environmental Quality is still responsible for advising the US president on environmental matters and coordinating governmental efforts with regard to the environment (The White House 2010). They also oversee the environmental impact assessment process called for by NEPA, previously discussed in Chapter 2. Remember that this process requires federal activities, or activities with federal funding, to perform an environmental impact assessment for public review before taking action.

In the United States, many state governments have taken up, at least partially, where the CEQ reports left off. In my home state of Iowa, the state Department of Natural Resources (IDNR) publishes a report entitled "State of the Environment" (IDNR 2010). The 2010 report examines eight indicators as a measure of our state's environmental health and quality. These are

1. Iowans Outdoors (hunting, fishing, camping, boating)
2. Land Protection (acres in conservation status)
3. Deer (population)
4. Game Birds (population)
5. Wildlife Diversity (presence of "rare" species)
6. Clean Air (particulate pollution, greenhouse gases)
7. Clean Lakes (water clarity, nutrient levels)
8. Water Quality (water quality index score)

This report touches on the ways that many Iowans interact with their local environment, from recreational opportunities to the quality of air that we breathe and water that we drink. Each section is a mixture of description and data trying to paint a recognizable picture that the public can comprehend. Examining this report over a series of years could inform us of how our environment is changing, hopefully for the better. It is also a report card on the performance of our state Department of Natural Resources. We have entrusted this specific part of our state government to watch over and protect our environment, and this report is a way of judging the quality and effectiveness of its efforts.

There are of course many other examples of governmental and scientific reporting on the state of the environment. A quick Internet search can soon overwhelm the interested reader with many possible avenues of available information and opinion. While these reports are generally very good in providing information, many of them stop short of recommending specific action, especially on an individual level. That is really not their function. Their task of merely informing is tough enough. They strive to portray the changing environment in a manner that many people will be able to understand. The value of these types of reports is in education and motivation. They set the stage for us to act upon. They provide a basic evaluation of the environment and suggest to us which issues are most critical at the time. The rest is up to us.

GREEN ACCOUNTING—ISEW (INDEX OF SUSTAINABLE ECONOMIC WELFARE)

Green accounting is a general term for the modification of traditional accounting practices to include the environmental impacts of our economic system. There are many proposed methods and indices for incorporating environmentalism within our traditional monetary accounting structure. I have chosen an index proposed by Daly and Cobb (Daly and Cobb 1989), the Index of Sustainable Economic Welfare (ISEW), to represent all the proposed green accounting measures.

I read Daly and Cobb's book, *For the Common Good*, when I was in graduate school, and found it very interesting. It opened my mind to thinking about how our traditional methods of quantifying economic progress may at times be at odds with our own welfare and self-interest. Daly and Cobb criticize our standard calculation of gross domestic product, or GDP, as insufficient for measuring the health and well-being of our economy. GDP is an estimation of the value of all the goods and services produced by a nation. Traditionally, increases in GDP are considered good by definition. The basic GDP calculation is as follows:

$$GDP = PC + INV + GOVT + (EXP - IMP)$$

where
PC	= Private consumption
INV	= Investment
GOVT	= Government spending
EXP	= Exports
IMP	= Imports

GDP is often thought to be related to our standard of living. If GDP is increasing, we must be growing wealthier and better off. The more we spend, the more we invest, the more our government spends and the more we export to other countries—all of these increase GDP. However, does this increase in GDP necessarily translate to increases in our social and environmental welfare? Are we better off if we increase our consumption of energy-intensive nonrecyclable consumer products that end up overflowing our landfills? Are we better off if we invest in highly automated new factories with relatively few workers to produce these energy-intensive nonrecyclable products? Are we better off if our government decides to expand the prison system by building expensive maximum-security institutions so that it can house those convicted of petty crimes? These examples may seem silly, but obviously there are many examples we could think of that call into question the association between GDP and quality of life.

From a strictly environmental point of view, GDP does not necessarily account for the environmental impacts of our economic system. Adding to the burden of environmental degradation or depleting natural resources does not always figure into the calculation. Government environmental regulation that requires private investment, and government spending in general, may capture some of these impacts. However, where government regulation and spending on environmental issues is lax or nonexistent, GDP will not reflect the environmental cost.

Green accounting attempts to remedy this situation by creating a new calculation that specifically includes some of these environmental factors. Daly and Cobb's ISEW includes subtracting the cost of environmental degradation and subtracting the depreciation of natural capital from the traditional GDP calculation. This is similar in concept to the first two pillars of practical environmentalism. The ISEW does not permit us to ignore the harm we inflict upon the environment or the wanton depletion of natural resources. In modifying GDP in this way, it attempts to portray a truer picture of changes to our standard of living, at least at the national scale. Incorporated within the ISEW is an economic measure of environmental performance. We could focus on this modification, essentially the difference between the GDP and ISEW, and use that as our measuring stick for the state of our environment.

SIMPLE MONETARY ECONOMICS

You might not think much of our standard economic measures as indicators of environmental impact, especially after the previous discussion of green accounting. However, there is a school of thought that equates environmental protection with economic progress. The reasoning goes something like this. Poor countries with underdeveloped economies do not have the economic power to invest in protecting their environment. Poor countries with small population may enjoy a healthy environment, but populous countries without the economic ability to make environmental investments may face a degrading environment. Wealthy nations, on the other hand, have the ability to require pollution control measures and other environmental safeguards. People who are wealthy enough to meet their more basic needs have the luxury to value a healthy environment as well.

There is much empirical evidence to support this logic. If you traveled the globe visiting countries near and far, which ones would enjoy the healthiest environments—the rich nations or the poor ones? If you traveled within your own country and visited the wealthiest suburbs and most exclusive enclaves, would you find rampant environmental degradation there? How would this compare to the poorest slums in your nation?

There is an argument that states that if you improve the wealth of the poor, you will eventually improve the environment. Poverty robs us of choice. With no discretionary income, we are forced to satisfy basic and fundamental human needs regardless of the environmental cost and degradation. How greatly immoral it would be to ask someone to go hungry, or without clothing, or without shelter in order to safeguard nature and her processes so that the rest of us can benefit from a healthy, vibrant environment.

Once basic needs are met, additional income can be invested to our choosing. When we do not fear hunger or cold, we can begin to think about the air we breathe and the water we drink. Our personal values can then enter into our purchasing decisions, and we can spend some of our excess income to improve our quality of life. Perhaps we would choose to purchase a simple yet efficient wood stove that heats our home and cooks our meals. No longer would the smoke from an open fire fill our dwelling and settle within our lungs. The more efficient stove would require less wood from the forest and allow us to leave more trees standing. It may also produce less particulate pollution, and our local air quality could improve not just for ourselves but for our neighbors as well. Maybe we would construct a better and more efficient home, one that consumes less energy and has some form of modern sanitation, "indoor plumbing," as it was known in the United States decades ago. Our physical, financial, and emotional security benefits, as does the environment. In many ways, improving the quality of life of those struggling with poverty can improve environmental quality, too. It is our nature to improve our quality of life. Improving the health of the environment at the same time is an added benefit.

The logic of economic prosperity leading directly to environmental benefit is not foolproof, however. Our personal notions of improved quality of life do not always align with environmental benefit. Trading up from a bicycle to an old gas-guzzling, blue-smoke-belching automobile would not be a win from an environmental degradation or resource conservation point of view. It would improve our mobility and perhaps our status, but with a cost to the natural world. Neither would trading up from an efficient cottage near our place of employment to a grand house in the suburbs with an hour-long commute each way be a step in the "right" direction from nature's point of view.

An environmentalist would probably prefer a green accounting index over existing simple monetary economic indicators as a measuring stick for environmental quality. The green accounting methods tend to close some of the environmental loopholes in our consumer-oriented economic systems.

Application of economic metrics as an environmental measuring stick also assumes that all environmental impacts can be valued in monetary terms. The units for GDP or for the Index for sustainable economic activity are money, be it dollars, pesos, euros or yen. While I agree that environmental impacts can be valued in monetary terms, I also question if this is appropriate and fair. For example, I am sure that someone could "calculate" the environmental impact of a species of fish becoming extinct. I am not sure that we would all agree upon this valuation. I am very sure that we would not fully understand the impact to the ecosystem where these fish once lived, or the impact to the societies that depended upon these fish for their livelihood. This extinction valuation is at least partially dependent upon a prediction of the future. How would other species adapt to the absence of the fish? How would coastal communities adapt? What would the environmental changes or impacts be of these adaptations? As discussed in Chapter 4, predicting the future is very tough and often very inaccurate.

In addition to the problem of valuation, both economic measures are far distant from us personally. They are national metrics and not designed to be applied at the individual level. In this regard they are similar to the "physical" metrics discussed previously. All six methods function better at the national level than at the individual level, for this is where the data reside in a form that is workable. As noted previously, it is difficult for us to individually calculate our carbon footprint to estimate the impact of our actions upon the earth's carrying capacity, or to figure out the life-cycle costs of all the products and services we consume. It is hard for us to digest the myriad government and scientific reports and know how our actions contribute to their findings. So also is it difficult for us to incorporate our individual environmental actions into national measures of economic activity, whether they are compiled using "green" or traditional accounting methods. All of these metrics—carbon footprint, carrying capacity, life-cycle costing, government and scientific reporting, green accounting methods, and simple monetary economics—suffer from an application problem. It is simply too difficult to apply these metrics to our individual decisions.

Practical environmentalism differs greatly from these six metrics. While it does produce a numerical score, this score is not a valuation of the health of the environmental itself but rather an indication of how "green" or "environmentally friendly" our action is. It is very difficult, perhaps nearly impossible, at the present time to represent the state or health of the environment in terms that are meaningful to an individual. Practical environmentalism circumvents this quagmire by ignoring it. Practical environmentalism depends upon the environmental intelligence, compassion, and integrity of the individual to drive its operation.

The metric of practical environmentalism, the pillars, is about direction and action. It easily accepts the imperfection of our knowledge and instead relies upon the goodness of our intention. Inaccuracies in data and information do not negate the value of the metric. While highly accurate information is preferred over less accurate information, the pillars function quite well with reasonably accurate information. A little research and understanding, coupled with a robust yet conservative framework, go a long way toward identifying actions with environmental benefit.

I certainly do not intend to be too critical of the six metrics presented in this chapter. I have found each one to be personally beneficial in improving my own understanding about how we impact the natural world. While I have struggled over incorporating these six concepts into my own decisions and actions, I am grateful for all those who created, discussed, and debated these concepts, for they have moved our environmental discourse forward. Practical environmentalism depends on these metrics to prod our thoughts and to commit our energies to a better understanding of our relationship with the environment. It is precisely this understanding that informs our judgment of the pillars, especially the first two pillars of environmental degradation and resource conservation. We need the information contained within the government and scientific reports. We need the wake-up calls from the environmental advocacy groups. We need to exercise our minds to work through the logic of carbon footprints, carrying capacities, life-cycle costing, and green accounting. All these things make the exercise of practical environmentalism that much easier.

Practical environmentalism seeks to leverage our knowledge and understanding into beneficial environmental action. It desires to stand on the intellectual shoulders of those who have come before us and get a better view of how we fit within the natural world. It wants a vantage point that allows us to see both nature and ourselves. It appreciates those who urge us to consider the impact of our actions as well as those who diligently search for truth and understanding amid the complexity of our world. Knowledge is truly powerful, but without action, environmental progress remains incomplete. The pillars seek to be the framework that makes the exercise of practical environmentalism easier, more robust, and ultimately successful.

REFERENCES

Sampson, A.W. 1923. *Range and Pasture Management*. New York: John Wiley & Sons.

Hardin, G. and Baden, J. 1977. *Managing the Commons*. San Francisco: W.H. Freeman.

Worldwatch Institute. 2010. http://www.worldwatch.org (accessed September 28, 2010).

United Nations Environment Programme (UNEP). 2010. http://www.unep.org (accessed September 28, 2010).

Council on Environmental Quality (CEQ). 2010. http://ceq.hss.doe.gov/ceq_reports/annual_environmental_quality_reports.html (accessed September 28, 2010).

The White House. 2010. Council on Environmental Quality. http://www.whitehouse.gov/administration/eop/ceq (accessed September 29, 2010).

Iowa Department of Natural Resources (IDNR). 2010. 2010 State of the Environment. http://www.iowadnr.gov/files/2010report.pdf (accessed September 29, 2010).

Daly, H.E. and Cobb, J.B. 1989. *For the Common Good*. Boston, MA: Beacon Press.

14 Some Final Thoughts

In many parts of the world, environmentalism is very "in" these days. It is hip, progressive, and politically correct. It is talked about in corporate boardrooms, as well as school classrooms. It has progressed beyond the attention-grabbing paramilitary tactics of antiwhaling, antilogging, and other antisomething groups to the point of common acceptance. Indeed, extreme environmentalism is very passé these days and seems quite unnecessary.

It has taken approximately 40 years to gain this level of acceptance and belief within our society. Much of this cultural shift occurred within the past 10 years. It seems like our society's views on the environment, its moral compass with respect to all things environmental, reached a tipping point very early in our 21st century and landed solidly on pro-environment footing.

If you are an environmentalist, and more and more of us profess that we are, then you should be relatively happy about this recent shift in attitude. It is a brand-new game these days, and the environment is getting the headlines and sharing the stage with all the other very important issues that are near and dear to us. If you pay attention to headlines, advertising campaigns, corporate annual reports, and even school curricula, being environmental is very middle of the road, very mainstream.

I used the term *tipping point* a couple of paragraphs previously in very deliberate fashion. The phrase is somewhat new to our common vocabulary and has gained a certain panache within environmental discourse. Tipping point implies a state where a relatively small action can induce a relatively large change in state. Malcolm Gladwell published a very interesting book about the concept (Gladwell 2000) with regard to sociological change. In the environmental arena, the term is often used in negative fashion to suggest that we are nearing a point of no return, a point where we have damaged the natural world to such an extent that we will not be able to halt the accelerating pace of environmental degradation or will we not be able to adapt to the reduction in natural resources that nature used to supply.

My use of the term was intended in a much more positive way. I used it to describe a change in humanity's thinking that is much more aligned with environmental awareness and consciousness. Somehow in the last decade or two, our modern societies have radically evolved from a seeming unconsciousness for the environment to a tacit acceptance that the environment requires rescue. What a huge shift in thinking and perspective! Sometime in our very recent past, environmentalism has gained enough momentum to induce a drastic change in the way people all around the world think about the environment.

I often wonder how such a great change could have occurred so rapidly. Perhaps it really has been a natural evolutionary shift in thinking, just one that occurred at an accelerated pace. Perhaps environmentalism struck a chord within our souls and our psyche that allowed a rapid acceptance. Perhaps the age of environmentalism fortuitously coincided with the beginnings of globalization and the explosion of computer and communications technology. Our world seems to turn a bit faster now and perhaps change happens faster now too.

I used the term because of the emotion that comes with it. Tipping point has a decidedly negative connotation within environmental discourse. It often promotes feelings of worry and anxiety. Indeed, part of the change in our environmental views is that we have come to worry about the environment and our place in the environment of the future. The worry and anxiety connected with the term highlights the emotion we naturally and often unconsciously feel when we think about environmental issues.

This state of worry, anxiety, and fear is often present within our modern environmentalism. It can be shallowly buried behind a green do-gooding façade, but it is still there. We can ignore it or pretend that it does not exist, but it is with us nonetheless. At its core, environmentalism is a fear-based science. If our environment were infinitely pliable and robust, we would not entertain one thought about our actions creating a negative impact. If all the soot-filled exhaust from the factory's smokestack was instantly vaporized and transformed into harmless breathable air upon its release, we would not have air quality regulations or pollution control equipment. If coal, oil, and gas were naturally regenerated just as fast as we consumed them, then energy conservation would be very close to ridiculous.

But these notions are indeed untrue. The pollution emanating from the smokestack does have the capacity to cause harm, and fossil fuels are not freely available. Excessive pollution can bring disease to our bodies. The sooty particles and acidic chemicals just might find their way from the smokestack to the air we breathe. They just might lodge deep within our lungs and impair our ability to transfer oxygen from the air to our bloodstream to satisfy the needs of our bodies. The scarcity of fuel that comes from overuse or even from manipulation of supply can cause shortages and drives prices beyond our reach. Our homes might grow cold when there is no fuel for the furnace. Our lives might become harder with the inconvenience and isolation that comes when there is no fuel to power our cars, buses, and trains. It is scary to think of the economic chaos that could ensue when energy supplies dwindle far below what our cities require. We shudder at the thought of more and more people competing for fewer and fewer resources.

For many of us, these fears are logically baseless. Their probability of occurrence is very low, far lower than many other issues that we should fear instead. Environmental regulations continue to be strengthened, and our scientific experts tell us that there are plenty of fossil fuels available for our use, hundreds of years worth. We should probably worry more about eating too much, drinking too much, or trying to talk on our cell phone while we drive. These things are far more likely to negatively impact our lives than the possibility of a complete melt down of the natural systems and resources that sustain us.

But this is not necessarily true for all of us. It is generally true for those who live in wealthy nations and benefit from a high standard of living. It is certainly less true for those who depend more directly upon Mother Nature for their daily survival. In this case, the fear of hunger, cold, or disease can be very real and very logical. In this case, recognizing one's dependence upon the environment is quite appropriate.

So whether our environmental fears are based in logic or not, I think that it is critically important to recognize that fear exists within our view of environmental issues. It may not be obvious or readily apparent, but our fear will be found if we dig deep enough. We may be so focused on data, computer projections, and scientific reasoning that the emotion of fear is driven deep into our subconscious. But it is there. It has to be there, or environmentalism itself has little basis. Recognizing the fear embedded within environmentalism is important because we tend to react to fear instead of responding to it. Fear often drives a fight or flight adrenaline-filled reaction. We impulsively act in order to protect ourselves. These reactions are immediate and usually with very little forethought.

Responses, on the other hand, are reasoned and more thoughtful. Practical environmentalism desires that we respond to environmental issues instead of reacting to them. The pillars are structured as they are for this very purpose. They are deliberately designed to slow down knee-jerk reactions to environmental calamities and transform them into reasoned responses. Creating a scorecard, defining a status quo, and evaluating each of the four pillars against this status quo, all require some thought and a little time to accomplish. The structure and process of the pillars help us to defuse highly charged emotions and, for a moment at least, lay aside the fear within us and constructively address the issues we face. Admittedly, our emotions do have a proper place in our decision making and, as is hopefully evident by now, the pillars allow and encourage their inclusion at the appropriate time.

The pillars are meant to help us with our choices. They help us combine the practical compromises we make every day with the environmental impacts of our actions to reach a conclusion, or chart a path that makes some sense. The pillars are necessary because of the way we typically approach environmentalism. As individuals we might care, we might believe, but we often do not think long enough. We tend to avoid the hard work and self-examination that is necessary to understand what is at stake with our choices. We must accept that our own personal choices can impact the environment. We must recognize that ill-conceived environmental programs can be detrimental to our welfare. In trying to save the planet, foolishly leaping forward without knowing where we might land could harm more than it helps.

When we consider our own personal choices, there are two main factors to consider that link environmentalism to our individual consumption patterns. The first is the direct environmental impact of the good or service itself. As mentioned in the discussion of life-cycle costing in the previous chapter, goods and services often come with hidden environmental costs. With a little foresight and good information, we can choose to minimize the environmental impacts of the things we buy.

The other factor, and at least equally important, is the environmental impacts we bring upon ourselves by choices of lifestyle. Our lifestyle often "forces" us to purchase certain types or quantities of goods and services in support of our chosen lifestyle. If we choose to drive a fast and powerful automobile, then we are bound to purchase the amount of gasoline to make it go. This is probably much more gasoline than would be required for a slower, less powerful, and more fuel-efficient vehicle. If we choose to live far from our place of employment, then we are bound to the transportation cost of our long commute and the environmental impacts that come along with it.

The pillars are structured to make it more obvious when these situations arise, but they do not do the work for us. We still need to do our homework. We still need to do our research. We still need to keep an open mind and consciously seek out viewpoints contradictory to our own to make sure that we have heard both sides of the issue. And with environmental issues, there are rarely only two sides to the story. The pillars suggest that there are at least four sides, with detailed arguments and subtle nuances within each one.

The pillars give us the structure to support our search for approaches that make sense. They give us the framework that ensures that individual concerns are included in the equation. Individual benefit is crucial because environmental progress is moving away from cleaning up obvious pollution and moving toward societal shifts in the basic ways that we live and work. This is big-issue environmentalism, but it still relies heavily upon individual behavior. We are quickly passing the point where environmental progress is cheap and easily hidden within the natural inflation in the cost of the things we purchase. Soon the effects of environmental progress could hit our wallets hard enough for us to really notice. Calls for environmental improvement are becoming synonymous with calls for us to change ourselves. We are no longer being asked to merely tidy up after ourselves; we are now called upon to save the world.

While I preach the virtues of practical environmentalism's focus on the individual, I know that I cannot exclude our collective activities and their impact on our planet. Our historical collection of individual actions summed up over time has yielded the "big issues." The big worrisome issues seem to grab most of the attention and most of the press, yet these are exactly the issues that we as individuals feel least able to influence. Practical environmentalism must be viable at this scale, even acknowledging that progress on the big issues is often difficult. We individually can do little to restore the Ogallala Aquifer, make radioactive waste less hazardous, or return our global climate to its preindustrial age balance. But we can pass judgment on programs and proposals that try to exert a positive influence on these big issues. We can use our judgment to select action that has the highest probability for long-term success. The 11 examples discussed in the big-issue chapters show a wide range in pillar scores and support the old adage that "there is more than one way to skin a cat." My apologies to PETA for this crude and insensitive reference. My family's pet of choice is the feline. We have two.

Returning for a moment to the fear behind environmentalism, there can be something good that comes from it. It is a rare case when there is only one side to any argument. As we recognize the deep-rooted fears of scarcity and want, we should also consider the fear associated with making mistakes. Students know this fear very well, and many experience it with every exam they suffer through. They double-check their math or reread the response to a question in hopes of recognizing and correcting mistakes before their teacher does. These students may have studied diligently and know the material thoroughly, yet they recognize the possibility of a silly oversight rendering their work incorrect.

In environmental matters mistakes usually occur out of ignorance rather than stupidity. And ignorance here is meant in the best possible terms. Most people who dare to try to better the environment are not stupid. On the contrary, environmental advocates are usually well educated and very committed to developing a deep understanding about the part of environmentalism that they are passionate about. But even those who study diligently will never know their material completely. The environment is far too complex to allow humanity a complete understanding of its workings. There are so many things we do not know. We do not know why the dinosaurs disappeared or why the glaciers came and went. We do not know why our earth has come to be in orbit around the sun or why the moon orbits around our earth. We really do not know why we sleep at night or why we dream about the things we do. We do not know what happens to us after we die.

Similarly, environmental mistakes do not arise from lack of effort or lack of good intention, but rather from lack of complete understanding. Often, we do not realize we have made an environmental mistake until years or decades have passed. Invasive species, contaminated groundwater, topsoil erosion, and perhaps even global climate change all suggest that environmental mistakes are easy to make. We must recognize that we possess only partial knowledge and that even with the best of intentions we are likely to make mistakes from time to time.

So what is wrong with that? There is an old saying that goes, "You cannot make an omelet without breaking some eggs," implying that mistakes come with the territory. There is no progress without some mistakes along the way too. So, is there really anything to fear with the current environmental credo? Is there anything wrong with assuming that no effort or cost is too great to protect our environment?

Practical environmentalism maintains that the answer to these questions is yes, certainly. Indeed, within these pages, we have seen how the pillars provide support for some environmental actions and discouragement for others. Fear of making mistakes can be turned to our advantage by encouraging our analyses to be a little better and a little more thorough. The pillars illustrate this advantage by grading some actions positively and others negatively, even when these actions have the same goal in mind. Let us briefly recap the examples used in this book to illustrate how we can use the pillars to help us make decisions with regard and respect for the environment.

Table 14.1 is organized by chapter. Recall that Chapter 9 introduced us to using the pillars with our personal decisions. Once again, this is really the heart of practical environmentalism. The four examples given in this chapter illustrated several ways that we could favorably impact the environment while also benefiting ourselves. It advocates individual action that aligns environmental benefit with personal benefit. None of the examples really involved much sacrifice, but you could easily come up with some scenarios that scored negatively on the pillars. I wanted to save the discussion of the sacrificial issues to Chapter 11, so I chose more positive examples here.

Chapter 10 was concerned with bigger issues. It was meant to show that the pillars could be useful at this scale as well. In Chapter 10, we saw the pillars land on the side of the Hetch Hetchy Dam, both in its initial construction, as well as prohibiting its demolition. In this particular case, pillar analysis suggests that removing the dam is a mistake we should avoid. The pillars also found in favor of efforts to limit concentrations of CFCs, SO_2, and NO_x in our atmosphere, and supported the institution of the national 55 mph speed limit. I would venture to guess that some environmentalists might take exception to the pillar analysis of Hetch Hetchy while supporting the pillar analysis of the ozone hole, cap, and trade for air pollution, and the 55 mph speed limit. I say this to emphasize that pillar analysis is not the same

TABLE 14.1

Examples of Pillar Analyses Used in This Book

Chapter	Issue	Score
9	Walk to work or school, Example #1	+7
9	Walk to work or school, Example #2	0
9	Purchase VW Golf TDI	+6
9	Compact fluorescent light bulbs	+3
10	Build the Hetch Hetchy Dam	+1
10	Remove the Hetch Hetchy Dam	−2
10	Ban CFCs	+1
10	Cap-and-trade sulfur dioxide emissions	+2
10	Institute 55 mph speed limit	0/+1
11	Yucca Mountain nuclear waste storage facility	−4
11	Ban irrigation from the Ogallala aquifer	−2
11	Cap-and-trade Ogallala water rights	0/+1
12	Sequester carbon dioxide	−5
12	Institute renewable energy portfolio standards	+4
12	Institute tax on carbon	+1

thing as environmental analysis. It includes environmental analysis, of course, but also reaches beyond as it tries to bring all four pillars into focus and ideally into alignment. Chapter 10 was our first real test of this paradigm in that we applied the pillars to familiar historical events to highlight the holistic nature of practical environmentalism. Practical environmentalism is indeed bigger than the environment it seeks to improve.

These bigger issues carried over into Chapters 11 and 12 too, and both chapters highlighted some potential mistakes and pitfalls that the pillars might help us avoid. Chapter 11 was highly focused on the negative scores associated with sacrificial environmental action but stressed how we can use the pillars to guide us toward less sacrificial options that still move us in a positive direction.

Chapter 12 is a critically important chapter simply due to all the attention directed to the single issue of global climate change. The three examples given in this chapter showed a wide range of pillar scores with one very negative, one very positive, and one near neutral. Our pillar analysis did not depend upon whether you believe the global warming theory or not. Nor did it require that we find the optimum solution to this global concern. Instead, the pillars showed their ability to differentiate between potential responses to this perceived threat. They offered us the opportunity to rationally compare actions that we might take to alleviate the climate change problem. They showed that there could be a big difference in proposed responses and tried to point out which responses may later turn out to be a mistake.

For all the "big" issues, practical environmentalism enables us to be better consumers of information. For the "little" issues, practical environmentalism can help us be better actual consumers. The products and services that we purchase, either directly from our own purse or through the expenditure of public monies, all have environmental impacts associated with them.

So, here is some advice to all who are confronted with environmental choices. If you are in a position of power, authority, and influence, remember that environmental issues are complex. Remember that there are at least four broad areas, four pillars, that are worthy of diligent reasoning and consideration. Remember that individuals matter and that environmental impacts, as well as social impacts, ultimately and inevitably trickle down to individuals.

If you are an individual facing important environmental choices, do not let others do all your thinking for you. Use the pillars, even in their most basic and brief form, to gauge how greatly the environmental choice will impact you. If the issue or impact is great, you can delve in further, become more knowledgeable and join the dialogue.

I truly hope that our environmentalism can be an inclusive dialogue, or at the very least a civil debate. Similar to it or not, and to greater or lesser extent, we are all in this together. We draw from the same pool of resources. We drink the same water and breathe the same air. My children and grandchildren will share the planet with yours. I know no one who favors pollution and environmental damage. When we disagree it is a matter of degree, a matter of approach rather than right or wrong, good versus evil.

Let us use our greatest resources, our collective intelligence, energy, and compassion, to forge a path forward to environmental progress and environmental health. Let us make reasoned, rational choices to improve the planet in which we live. Let us not forget those whose voices are quiet and meek as we enter into grand governmental policies, programs, and plans. Let us always be humble and accept the finiteness of our knowledge and the uncertainty of our predictions. Let us take actions that can be managed and modified as we learn from our future. Let us walk gingerly into the future, valuing our environment, as well as our neighbor. Let us truly make this world better for our children.

So, now that you are armed with the knowledge of practical environmentalism and the four pillars, you might wonder, "What specifically you should I do to benefit the environment?" Let us address this question by completing the top 10 list of environmental actions that I promised you in the introductory chapter of this book.

Since we have spent a good amount of time and a fair amount of pages applying the pillars to specific environmental actions, let us use the examples summarized in Table 14.1 as the starting point for our top 10 list. The best scoring examples in Chapters 9 through 12 were

1. Walk to work or school, +7
2. Purchase a very efficient car, +6
3. Institute renewable energy portfolio standards, +4
4. Use compact fluorescent light bulbs, +3
5. Cap and trade for sulfur dioxide emissions, +2

There were only five examples that scored at least a +2 on the pillars, the minimum I would consider to represent a positive step in the march for environmental betterment. A +1 score is a little too close to neutrality for me to set it apart among the best environmental actions from any group of possibilities. Three of these top five are examples of individual actions. The other two are government programs, cap and trade at the federal level, and renewable energy portfolio standards proposed at the federal level but currently active in some states.

This list, however, is halfway short of the promised top 10. Even considering that 15 examples were presented in Chapters 9 through 12, we only discovered 5 that were good enough to make the cut. Of course, this book was not intended to merely be a checklist of "100 Green Things" for you to do to save the planet. Practical environmentalism is not about prescribing a set of beliefs for the pilgrim to accept in order to gain entry into the company of believers. Instead, this book was intended to present a framework and methodology to allow each of us to choose actions that will result in some measure of environmental improvement. The list of beneficial actions is dependent upon you, and hopefully at this point, it should come as no surprise that there are many environmental ideas that should not make your list at all.

So, let me finish this top 10 list with five suggestions based on my own opinions and preferences. While I hesitate slightly to do this in fear of my biases becoming yours, I trust by now that you understand how the pillars are supposed to function and that you are quite capable of creating your own ideas for action and scoring them as it suits you. Let us consider these additional five as a starting point for your own investigation into what environmentally friendly actions are best suited to your personal situation.

1. Use a clothesline to dry your clothes
2. Purchase a water filter to make your own drinking water from tap water
3. Patronize local merchants
4. Watch your home's thermostat closely and take advantage of fans in the summer and sweaters in the winter
5. Purchase products with a goal of eventually recycling them and their packaging

Pretty boring stuff, huh? There are no windmills, hydrogen-powered vehicles, organic home gardens, or compost piles in sight. Now if we examine the top 10 environmental actions, we see that eight of them are individual decisions and activities. All five suggestions that I added to the list were actionable personally, five little steps toward environmental improvement. This bias toward individual action is very much in keeping with the principles of practical environmentalism. Concentrating on our own actions means that we can easily effect change. It is almost entirely within our control. We can choose activities that align most or all of the pillars in a positive direction. We can create win-win scenarios. Perhaps individually we do not save the planet in this way but collectively over time we just might. Does not it feel good to take control over our environmental ethics and be able to make real changes within our own lives and know that the environment is better off because of our actions? Is not that better than worrying about the "big issues" over which we have little control? And is not it nice that we can improve the environment and better ourselves at the same time?

I chose these specific five additions for two main reasons. Number one, they are fairly generic. You could do some or all of these suggestions whether you live in a house, an apartment, or even a dorm room. They work equally well in the city or on the farm. The second reason for their selection is that naturally I expect them all to score well on the pillars, particularly the environmental degradation and resource conservation pillars. These five suggestions tend to minimize energy use that also reduces pollution from burning fuels to generate energy. This linkage is so strong that it often makes sense to seek out resource conservation first and reap the rewards of environmental improvement that follow. I think these five suggestions would also score well on economic progress and personal benefit pillars too. Saving energy usually means saving money, and is not it nice to smell the fresh sent of air-dried clothes and feel a gentle breeze on a warm day?

As we come to the end of this book, let me offer you a challenge. Since practical environmentalism is really about your personal choices and actions, run these last five suggestions through your own pillar analysis and see what you come up with. I would be surprised if you found any of them to be less than positive.

True environmentalism must benefit both the group and the individual. Practical environmentalism exists in finding ways to make this happen. It has to do with the art of the possible.

REFERENCES

Gladwell. M. 2000. *The Tipping Point: How Little Things Can Make a Big Difference.* Little, Brown.

Index

.